# Research Diary

## *2018*

Aleks Kleyn

Aleks_Kleyn@MailAPS.org
http://AleksKleyn.dyndns-home.com:4080/
http://sites.google.com/site/AleksKleyn/
http://arxiv.org/a/kleyn_a_1
http://AleksKleyn.blogspot.com/

ABSTRACT. In this book, I considered problems which I faced during 2018:
- general covariance principle, role of coordinate and coordinate-free representations of geometric object, description of experiment to test unholonomity of coordinates in gravitational field;
- Cauchy-Riemann-Fueter equation, attempt to write down the equation in coordinate-free form and analysis of how differential forms look in module over non commutative $D$-algebra;
- linear algebra, study of basis of module of linear maps, system of linear equations in quaternion algebra;
- division of polynomials in non commutative algebra, description of the set of quotients;
- example of noncommutative sum in affine space.

Kindle Direct Publishing

ISBN: 1-0921-2370-9

ISBN-13: 978-1092123709

Translated from Russian
Дневник поиска
2018
Александр Клейн

# Contents

Chapter 1. Preface ............................................. 6
1.1. Why do I write these notes? ............................... 6
1.2. Craftiness of Partial Derivative .......................... 6
1.3. Cauchy-Riemann-Fueter Equation ............................ 7

Chapter 2. 2018 ............................................... 11
2.1. Eigenvalues .............................................. 11
    February 18 ............................................ 11
    February 19 ............................................ 11
2.2. Geometric Object ......................................... 11
    February 19 ............................................ 11
2.3. General Covariance Principle ............................. 12
    February 20 ............................................ 12
    April 25 ............................................... 13
2.4. Cauchy-Riemann-Fueter Equation ........................... 14
    February 20 ............................................ 14
    February 21 ............................................ 14
    February 22 ............................................ 15
    February 24 ............................................ 16
    February 25 ............................................ 17
    February 26 ............................................ 18
    February 28 ............................................ 21
    March 2 ................................................ 23
    March 3 ................................................ 23
    March 4 ................................................ 24
    March 5 ................................................ 25
    March 6 ................................................ 26
    March 7 ................................................ 27
2.5. $D$-Algebra .............................................. 28
    March 16 ............................................... 28
    May 6 .................................................. 29
2.6. Anholonomic Time ......................................... 31
    March 17 ............................................... 31
    March 18 ............................................... 32
    March 19 ............................................... 32
    March 20 ............................................... 33

| | March 21 | 34 |
|---|---|---|
| | March 22 | 36 |
| 2.6.1. | File NonHolonom.aspx | 37 |
| 2.6.2. | File NonHolonom.aspx.cs | 38 |
| 2.7. | Linear Map of Octonion Algebra | 40 |
| | May 20 | 40 |
| 2.8. | Structure of Linear Map | 40 |
| | June 13: Basis of $A \otimes A$-module $\mathcal{L}(D; A \to A)$ | 40 |
| | June 14 | 41 |
| | June 15 | 41 |
| | June 16: Maps of Conjugation | 42 |
| | June 19 | 42 |
| 2.9. | Non Commutative Sum | 42 |
| | June 16 | 42 |
| | November 5 | 45 |
| | November 8 | 46 |
| | November 10 | 47 |
| | November 13 | 47 |
| 2.10. | Free Representation | 47 |
| | August 9 | 47 |
| | August 10 | 49 |
| | August 11 | 49 |
| | August 13 | 50 |
| | August 16 | 50 |
| | September 23 | 51 |
| | September 30 | 51 |
| | November 27 | 52 |
| | November 29 | 53 |
| 2.11. | System of Linear Equations | 54 |
| | August 18: Quaternion Algebra | 54 |
| | August 19 | 55 |
| | August 20 | 56 |
| | August 21 | 58 |
| | August 22 | 59 |
| | August 23 | 60 |
| | August 25 | 60 |
| 2.12. | Division of polynomials | 63 |
| | October 6 | 63 |
| | October 7 | 64 |
| | October 8 | 65 |
| | October 10 | 66 |
| | October 11 | 68 |
| References | | 74 |

Index . . . . . . . . . . . . . . . . . . . . . . . . . . . . . . 76

Special Symbols and Notations . . . . . . . . . . . . . . . . . . 77

# CHAPTER 1

# Preface

## 1.1. Why do I write these notes?

When I was young, I taught at a junior math school. I had a tradition to offer a labyrinth problem during the first lecture and solve it with students in the style of Polya: I ask questions; students try to answer. On the one hand, trial and error method resembles wandering in a maze. On the other hand, to solve the problem it is important to raise the question properly.

However, the maze changes over time.

The goal of these notes, which I have been doing for many years, is to save problems and solutions which I met on my way. I do not exclude that these notes may be interesting to my reader as well. Papers and books which we write are complete work. We edit them. But research is a deeply emotional process. Any ideas that are not directly related to the text may appear outside the scope of the text and will be lost forever.

When I started to use text editor to prepare papers, I have not been doing my research diary for some time. When I saw hard problem, I experimented immediately in the text of the paper. If I was not satisfied by the answer, I ruthlessly deleted everything related to experiment. Later I realized that this was my mistake; and I resumed to keep notes in diary.

## 1.2. Craftiness of Partial Derivative

At the beginning of this year, I decided to write short text [17] where I assembled only definitions from single variable calculus and theorems without prove. I believe that such text allows the reader to see the theory from the height of bird flight without being distracted by details which overshadow general construction. I suppose to write similar texts in the future.

When the text was ready, I was musing. In the papers [7, 9], the text which I dedicated to multivariable calculus was preliminary. I needed this text to understand the structure of linear map of vector space over non commutative division algebra. I still have to write a paper dedicated to module over non commutative algebra before I will write paper dedicated to multivariable calculus. However, if I used only definitions and statements of theorems, why not sketch the future theory.

Here I met a surprise which threw me into amazement. Jacobian matrix of a map did not depend on whether I consider the left or right modules.

I will explain using following example. Let $\overline{\overline{e}} = (e_1, ..., e_n)$ be a basis of module $V$ over $D$-algebra $A$. Then we can consider the map of $V$-number as map of $n$ $A$-numbers

$$f(x = x^1 e_1 + ... + x^n e_n) = f_1(x_1, ..., x_n)$$

or

$$f(x = e_1 x^1 + ... + e_n x^n) = f_1(x_1, ..., x_n)$$

However, it does not matter for the map $f$ what kind of module we consider: left or right.

This result did not contradict to my knowledge about linear maps. However 10 year ago, I had written the paper [7] where I considered the question about solving of a system of linear equations over non commutative algebra. The set of maps which I considered in this paper was small subset of the set of linear maps. This was the most difficult period in my research.

After a week of intense thought, I decided to rewrite the paper [7]... I did not make this. Considering diagram of representations describing module over $D$-algebra, I realized that map in the paper [7] is reduced morphism of diagram of representations.

I was not lazy and looked into the book [3]. Lang also considered such maps. He called these maps homomorphisms of module. It is naturally. To study module over $D$-algebra we need to study linear maps and homomorphisms. We need linear maps in calculus. We need homomorphisms to study transformations of coordinates of vector when we change a basis.

I realized that it is time to write book dedicated to linear algebra; I expect that this book will cover a wide range of topics: linear maps and homomorphisms, eigenvalues, systems of linear equations. At the same time, I have accumulated some interesting results related to the theorey of quadratic equations, I saw relationship between polylinear map and polynomials. The result was a project to write textbook dedicated to non commutative algebra.

In the beginning, I did not know what kind of this book will be and how large it will be. But an accident helped me. I decided write in the first place a small paper dedicated to linear maps of quaternion algebra and octonion algebra. But if I consider linear maps, why not to solve a couple of systems of linear equations.

I do not write about problems which I met (see, for instance, the section 2.11). But the paper overgrown important details. At some point of time, I realized that paper becomes volume one of future project.

## 1.3. Cauchy-Riemann-Fueter Equation

I started research in calculus over non commutative Banach $D$-algebra on 2008. My knowledge about the structure of a linear map was not clear and I was looking specific examples where I could see linear map. I have not seen a better candidate than calculus.

The abundance of candidates for the role of derivative justified my choice. It took me about two years before I was able to see the structure of linear map.

Quaternion algebra is simplest example of non commutative Banach algebra. I was wondering what other mathematicians write about differentiation of a map of quaternion algebra. Research was based on consideration of derivative as Jacobian matrix of the map. In particular, my attention was attracted by Cauchy-Riemann-Fueter equation. This equation was like Cauchy-Riemann equation in complex field. But in my opinion it was not enough.

Function of a complex variable which satisfies to Cauchy-Riemann equation

$$(1.3.1) \qquad \begin{aligned} \frac{\partial f^1}{\partial x^0} &= -\frac{\partial f^0}{\partial x^1} \\ \frac{\partial f^0}{\partial x^0} &= \frac{\partial f^1}{\partial x^1} \end{aligned}$$

is called analytical. The equation $(1.3.1)$ is equivalent to to the equation

$$(1.3.2) \qquad \frac{\partial f}{\partial x^0} + i\frac{\partial f}{\partial x^1} = 0$$

Analytic functions are infinitely differentiable. The integration theory of analytic functions is extremely interesting and rich.

We define complex conjugation by the equality

$$(1.3.3) \qquad a = a^0 + a^1 i \Rightarrow a^* = a^0 - a^1 i$$

From the equality $(1.3.3)$, it follows that

$$(1.3.4) \qquad a^0 = \frac{1}{2}(a + a^*) \qquad a^1 = -\frac{i}{2}(a - a^*)$$

From the system of differential equations $(1.3.1)$ and from the equality $(1.3.4)$, it follows that

$$(1.3.5) \qquad \begin{aligned} \frac{\partial f}{\partial x^*} &= \frac{\partial f^0}{\partial x^*} + \frac{\partial f^1}{\partial x^*} i \\ &= \frac{\partial f^0}{\partial x^0}\frac{\partial x^0}{\partial x^*} + \frac{\partial f^0}{\partial x^1}\frac{\partial x^1}{\partial x^*} + \left(\frac{\partial f^1}{\partial x^0}\frac{\partial x^0}{\partial x^*} + \frac{\partial f^1}{\partial x^1}\frac{\partial x^1}{\partial x^*}\right)i \\ &= \frac{1}{2}\left(\frac{\partial f^0}{\partial x^0} + \frac{\partial f^0}{\partial x^1}i + \left(\frac{\partial f^1}{\partial x^0} + \frac{\partial f^1}{\partial x^1}i\right)i\right) \\ &= \frac{1}{2}\left(\frac{\partial f^0}{\partial x^0} + \frac{\partial f^0}{\partial x^1}i + \frac{\partial f^1}{\partial x^0}i - \frac{\partial f^1}{\partial x^1}\right) \\ &= 0 \end{aligned}$$

The equality $(1.3.5)$ gives us opportunity to consider the set of maps which do not depend on complex conjugated variable.

## 1.3. Cauchy-Riemann-Fueter Equation

Similarly, in quaternion algebra, the map $f$ is called regular, if the map $f$ satisfies to Cauchy-Riemann-Fueter equation[1.1]

(1.3.6) $$\frac{\partial f}{\partial x^0} + i\frac{\partial f}{\partial x^1} + j\frac{\partial f}{\partial x^2} + k\frac{\partial f}{\partial x^3} = 0$$

We can consider quaternion algebra as vector space over complex field. Let

$$q = q^0 + jq^1$$
$$f(q) = f^0(q) + jf^1(q)$$

Then the map $f$ satisfies to the system of differential equations[1.2]

(1.3.7) $$\frac{\partial f^0}{\partial q^{0*}} = \frac{\partial f^1}{\partial q^{1*}}$$
$$\frac{\partial f^0}{\partial q^1} = -\frac{\partial f^1}{\partial q^0}$$

where $q^{0*}$, $q^{1*}$ are complex conjugated variables.

Similarity of equations (1.3.2) and (1.3.6), as well as of systems of differential equations (1.3.1) and (1.3.7), says that there is a similarity between complex field and quaternion algebra. But how deep is this similarity?

First of all, there is no analogue of the equation (1.3.5) in quaternion algebra. Consider this question in more detail. I continue to consider quaternion algebra as vector space over complex field. Conjugation in quaternion algebra has the following form[1.3]

(1.3.8) $$q^* = q^{0*} - jq^1$$

However, we cannot find unique expression for $q^0$ and $q^1$ using $q$ and $q^*$ as we found in complex field, because, in quaternion algebra, $q^*$ is linear function of $q$.

The possibility of the existence of the set of functions which satisfy to the equation (1.3.5), is due to the fact that the kernel of complex field is larger than real field. To better understand this problem, I wrote the paper [12]. Later, thanks to this paper, I realized that the set of linear maps of quaternion algebra is left algebra of dimension 4 over quaternion algebra.

In the process of research, my understanding of the role of Cauchy-Riemann-Fueter equation changed. The paper [21] played a significant role in this. It was important to me that this equation is consequence of theory of differential forms in quaternion algebra. However, it took me a while to find out in detail how to form this equation using theory of differential form.

As a result of this study I understood the structure of differential forms in module over non commutative algebra. I will start from afar.

In 2007, when I studied vector space over division algebra, I asked myself how an equation of plane might look like. The construction of the plane is the

---

[1.1]See the proposition [21]-3 on the page 209 and equation [21]-(3.10).

[1.2]See the proposition [21]-3 on the page 209 and equation [21]-(3.11).

[1.3]This looks a little weird. However, we see conjugation of quaternion algebra in left side and conjugation in complex field in right side.

same as in vector space over field. I choose $m$ vectors $v_1$, ..., $v_m$, which are basis in the plane and write down the condition that any vector of the plane is linear combination of vectors $v_1$, ..., $v_m$.

To solve this problem, it is enough to write the matrix, whose columns correspond to vectors $v_1$, ..., $v_m$, $v$, where $v$ is any vector of plane. Then write the condition that this matrix has rank $m$. In case of vector space over division algebra, I will get answer on the language of quasideterminants. However, what seems clear today surprised me 10 years ago.

I will get similar answer for external product of vectors and, therefore, for any differential form.

# CHAPTER 2

# 2018

## 2.1. Eigenvalues

**February 18.** There is one topic in non commutative algebra that I did not consider in papers. I am writing about eigenvalues of matrix.

Let $A$ be algebra over commutative ring $D$. Consider two forms of matrices.

I will start with a matrix $B$ whose elements are $A$-numbers. Formally, we speak about solution of equation
$$\det(B - \lambda E) = 0$$
Considering that instead of the determinant, I must use quasideterminant, I see here a system of equations. I will return to this question later.

However, the eigenvalues arise in the process of solving a certain problem. One of such problems arises when linear map maps a vector $a$ to vector $b$ which is parallel to vector $a$. However a matrix of $A$-numbers is particular form of limear map. In general, entries of matrix of linear map are $A \otimes A$-numbers.

**February 19.** So, I am looking for eigenvalues like $a_s \otimes b_s$, $a_s$, $b_s \in A$. At first glance this looks unusual. The set of vectors generated by vector $v \in V$ is $av$ in left $A$-module and $va$ in right $A$-module. Therefore, the set of vectors parallel to the vector $v \in V$ is larger than the set of vectors generated by vector $v$.

However, there is one more argument in favor of choosing a tensor as eigenvalue. I have in mind differential equation. I recall that the derivative of exponent has form
$$\frac{de^{a \circ x}}{dx} = e^{a \circ x} a + a e^{a \circ x}$$
However, this analogy is not so obvious.

The difference between matrix of $A$-numbers and matrix of $A \otimes A$-numbers is not limited to the problem of eigenvalues. I will consider another problem in section 2.2.

## 2.2. Geometric Object

**February 19.** There exist two complimentary groups in vector space, namely, they are group of active transformations and group of passive transformations.

Passive transformation maps one basis into another. To specify a passive transformation, it is sufficient to write the matrix of coordinates of new basis relative new one. This is matrix of $A$-numbers.

Passive transformation is responsible for transformation of coordinates of geometrical object when we consider map of module. Usually this map is linear map. Or this map is derivative of any map if we consider differential geometry. However coordinates of linear map are $A \otimes A$-numbers.

STATEMENT 2.2.1. *It turns out that, in a module over non commutative D-algebra for a given geometric object, transformation of coordinates depends on the choice of basis.* $\odot$

To what extent is the statement 2.2.1 true?

## 2.3. General Covariance Principle

**February 20.** Ideas in this subsection is a continuation of the statement 2.2.1.

General covariance principle is cornerstone of general relativity. My research in algebra and geometry suggests that this principle can be formulated in different mathematical theories. To understand how it works, I study passive and active transformations of basis manifold, as well I study transformations of coordinates of geometric object when the basis changes.

It may give the impression that I want to state a theory based on coordinate representation. However this is not so.[2.1] There is only one goal why I study coordinate transformation. Since I know how coordinates change when I change basis or reference frame, I can write expression which does not dependent on the basis. Unfortunately, even today I can see physical theory which is good in specific reference frame. Experiment supports this theory. But what will happens, if an observer will move with high speed? Will the observer see the same experiment?

I admit that speed of light may be variable in new theory. But I wonder when I hear that somebody discovered violation of Lorentz transformation in this theory.

Lorentz transformations are defined within the framework of special and general relativity. If experiment in general relativity show conditions when Lorentz transformations does not work, then this is very important achievement, because we will see boundaries of general relativity.

However, new theory has different geometry. Lorentz transformations in the new theory are not defined. We need to start from scratch the study of general covariance in new theory. We need to define a reference frame; in other words, we need define a basis with respect to which measurements are performed or we need to define measurement tool. We need to define transformation from one reference frame to another reference frame.

And only then we can say whether general covariance persists in new theory. If general covariance does not persist, then why it does not.

---

[2.1] My research in non commutative algebra shows that I try to work without coordinates.

There is one more reason why we need to know law of coordinate transformation. Any mathematical theory that claims to be a physical, sooner or later, must reach the experiment. We have to choose measurement tool and to predict what results these devices will show.

The task to determine the reference frame and law of transformation of geometrical object is not simple task. For instance, I tried to consider this problem for tangent space in Finsler geometry [14]. However transformations which I and Alexandre Laugier found do not generate group. It is possible that our definition of reference frame was not good.

The statement 2.2.1 created new concern as well.

**April 25.** Below I put fragment from my letter to Arbaaz Mahmood on Research Gate.

From development of physics during XX century, you may learn that some fundamental law may change during few years; however they keep the same time some kind of symmetry which is important in understanding of world.

Here you need to focus your attention.

Now I will tell you thing that initially will be shocking to you. The mass is not physical reality; it is math abstraction which Newton introduced to express his second law. (I myself was shocked when learned that wave is not physical phenomenon, but math abstraction which describes movement of energy through space).

There is one more math abstraction that we need to consider as well. The definition of space is just geometry which we need to use as language which describes physical law. I do not go deep into past. But I think Newton was first who introduce concept of time and space, namely absolute time and absolute space. He was needing these definitions to develop his mechanics.

The first request was symmetry of time and symmetry of space. There is no selected point in time and there is no selected point in space. There is no selected direction in space. If you go to Lagrange formalism first request generates definition of energy and energy conservation law. Second request generates definition of momentum and momentum conservation law. Third request generates definition of angular momentum and angular momentum conservation law.

However XX century created new reality. Maxwell law demanded that we need to join space and time into one reality, namely space time. It was done by Einstein and Minkowsky. New geometry created new definition. Symmetry of space time required not two separated conservation laws, but one; namely, this symmetry required momentum energy symmetry law.

The definition of mass came similar evolution. It is true that we observe mass conservation law in chemical reaction. But chemical reaction is only one specific type of physical phenomenon. Lorentz transformation required that we had to join energy and mass into single definition. Physics of XX century does not know separate energy conservation law or mass conservation law. But we know mass energy conservation law. This conservation law helped us to discover some particles like neutrino.

## 2.4. Cauchy-Riemann-Fueter Equation

**February 20.** I again return to Cauchy-Riemann-Fueter equation. Fueter equation follows from the equation

$$d(dq \wedge dq\, f) = Dq\, f' \tag{2.4.1}$$

where

$$Dq = dx^1 \wedge dx^2 \wedge dx^3 - i dx^0 \wedge dx^2 \wedge dx^3 - j dx^0 \wedge dx^3 \wedge dx^1 - k dx^0 \wedge dx^1 \wedge dx^2 \tag{2.4.2}$$

It gives me no rest, why the mapping $f$ is written either on the left or on the right of the differential form.

Repeating the consideration of professor Sudbery, I noticed that there is the derivative $f'$ in right side of the equation (2.4.1). However $f' \in H$ for Sudbery. And $f' \in H \otimes H$ for me. Maybe this is the solution?

It is only necessary to understand how to write it properly.

Let me define action for differential form $\omega$ and tensor $a \otimes b$

$$(a \otimes b) \circ \omega = a\omega b$$

Then I can try

$$(f \otimes 1) \circ (dq \wedge dq) = \frac{df}{dh} \circ Dq \tag{2.4.3}$$

$$(1 \otimes f) \circ (dq \wedge dq) = \frac{df}{dh} \circ Dq \tag{2.4.4}$$

It is good time to recall the equality

$$\frac{df}{dx} = -\frac{1}{2}\left(\frac{\partial f}{\partial x^0} + \frac{\partial f}{\partial x^1}i + \frac{\partial f}{\partial x^2}j + \frac{\partial f}{\partial x^3}k\right) \circ E$$
$$+ \frac{1}{2}\left(\frac{\partial f}{\partial x^0} + \frac{\partial f}{\partial x^1}i\right) \circ I + \frac{1}{2}\left(\frac{\partial f}{\partial x^0} + \frac{\partial f}{\partial x^2}j\right) \circ J$$
$$+ \frac{1}{2}\left(\frac{\partial f}{\partial x^0} + \frac{\partial f}{\partial x^3}k\right) \circ K$$

**February 21.** I will start from the statement

$$\frac{d}{dx}((dq \wedge dq)f) = (dq \wedge dq)\frac{df}{dx} \tag{2.4.5}$$

It is not very convenient that differential and exterior differential use the same letter $d$. To simplify calculations, I introduce differential form $\omega$

$$\omega \circ (dq, dq) = dq \wedge dq$$

Let play a little. Let

$$\omega = 1 \otimes_1 1 \otimes_2 1 - 1 \otimes_2 1 \otimes_1 1 \tag{2.4.6}$$

If
$$dq = dx^0 + idx^1 + jdx^2 + kdx^3$$
then we get ... Then we get nonsense.

Of course, this is nonsense; because I perceive $dq$ as $H$-number.

**February 22.** $dq$ is the differential form

(2.4.7) $$dq = 1 \otimes 1$$

According to the theorem [16]-8.1.3, the equality

(2.4.8) $$dq \wedge dq = [(1 \otimes 1) \underline{\otimes} (1 \otimes 1)] = \frac{1}{2}(1 \otimes_1 1 \otimes_2 1 - 1 \otimes_2 1 \otimes_1 1)$$

follows from the equality (2.4.7). In particular, the equality

(2.4.9)
$$(dq \wedge dq) \circ (a, b) = \frac{1}{2}(ab - ba)$$
$$= \frac{1}{2}((a^2 b^3 - a^3 b^2)i + (a^3 b^1 - a^1 b^3)j + (a^1 b^2 - a^2 b^1)k)$$
$$= (a^2 \wedge b^3)i + (a^3 \wedge b^1)j + (a^1 \wedge b^2)k$$

follows from the equality (2.4.8).

This is good. Just small difference in format of notation.

Now I can write either
$$\frac{d}{dx}((dq \wedge dq)f) = (dq \wedge dq)\underline{\otimes}\frac{df}{dx}$$
or
$$\frac{d}{dx}((dq \wedge dq)f) == \frac{d}{dx}(1 \otimes_1 1 \otimes_2 f - 1 \otimes_2 1 \otimes_1 f)$$
I will get the same answer in both cases.

$$d((dq \wedge dq)f) = \frac{1}{6}\left(1 \otimes_1 1 \otimes_2 \frac{d_{s \cdot 0}f}{dx} \otimes_3 \frac{d_{s \cdot 0}f}{dx} - 1 \otimes_2 1 \otimes_1 \frac{d_{s \cdot 0}f}{dx} \otimes_3 \frac{d_{s \cdot 0}f}{dx}\right.$$
$$+ 1 \otimes_2 1 \otimes_3 \frac{d_{s \cdot 0}f}{dx} \otimes_1 \frac{d_{s \cdot 0}f}{dx} - 1 \otimes_3 1 \otimes_2 \frac{d_{s \cdot 0}f}{dx} \otimes_1 \frac{d_{s \cdot 0}f}{dx}$$
$$\left. + 1 \otimes_3 1 \otimes_1 \frac{d_{s \cdot 0}f}{dx} \otimes_2 \frac{d_{s \cdot 0}f}{dx} - 1 \otimes_1 1 \otimes_3 \frac{d_{s \cdot 0}f}{dx} \otimes_2 \frac{d_{s \cdot 0}f}{dx}\right)$$

I see real problem here. $Dq\, f'$ is written on the right side. But $Dq$ itself is map of 3 variables. What I can do with $f'$? $f'$ also should bring one differential.

Assume for short time that $f'$ is a number. Then I will get

$$d((dq \wedge dq)f) \circ (a, b, c) = \frac{1}{6}\left(ab\frac{df}{dx}c - ba\frac{df}{dx}c + bc\frac{df}{dx}a\right.$$
$$\left. -cb\frac{df}{dx}a + ca\frac{df}{dx}b - ac\frac{df}{dx}b\right)$$

The following equality from [21] will be useful for me

$$Dq \circ (a,b,c) = \frac{1}{2}(ca^*b - ba^*c)$$
$$= \frac{1}{4}(b(a + iai + jaj + kak)c - c(a + iai + jaj + kak)b)$$

(2.4.10)
$$= \frac{1}{4}(1 \otimes_2 1 \otimes_1 1 \otimes_3 1 + 1 \otimes_2 i \otimes_1 i \otimes_3 1$$
$$+ 1 \otimes_2 j \otimes_1 j \otimes_3 1 + 1 \otimes_2 k \otimes_1 k \otimes_3 1$$
$$- 1 \otimes_3 1 \otimes_1 1 \otimes_2 1 - 1 \otimes_3 i \otimes_1 i \otimes_2 1$$
$$- 1 \otimes_3 j \otimes_1 j \otimes_2 1 - 1 \otimes_3 k \otimes_1 k \otimes_2 1) \circ (a,b,c)$$

The equality

(2.4.11)
$$Dq = \frac{1}{4}(1 \otimes_2 1 \otimes_1 1 \otimes_3 1 + 1 \otimes_2 i \otimes_1 i \otimes_3 1$$
$$+ 1 \otimes_2 j \otimes_1 j \otimes_3 1 + 1 \otimes_2 k \otimes_1 k \otimes_3 1$$
$$- 1 \otimes_3 1 \otimes_1 1 \otimes_2 1 - 1 \otimes_3 i \otimes_1 i \otimes_2 1$$
$$- 1 \otimes_3 j \otimes_1 j \otimes_2 1 - 1 \otimes_3 k \otimes_1 k \otimes_2 1)$$

follows from the equality (2.4.10). Let

$$d((dq \wedge dq)f) = \frac{1}{6}\left(1 \otimes_1 1 \otimes_2 1 \otimes_3 \frac{df}{dx} - 1 \otimes_2 1 \otimes_1 1 \otimes_3 \frac{df}{dx}\right.$$
$$+ 1 \otimes_2 1 \otimes_3 1 \otimes_1 \frac{df}{dx} - 1 \otimes_3 1 \otimes_2 1 \otimes_1 \frac{df}{dx}$$
$$\left.+ 1 \otimes_3 1 \otimes_1 1 \otimes_2 \frac{df}{dx} - 1 \otimes_1 1 \otimes_3 1 \otimes_2 \frac{df}{dx}\right)$$

Then I should get

(2.4.12)
$$\frac{1}{4}(1 \otimes_2 1 \otimes_1 1 \otimes_3 1 + 1 \otimes_2 i \otimes_1 i \otimes_3 f'$$
$$+ 1 \otimes_2 j \otimes_1 j \otimes_3 1 + 1 \otimes_2 k \otimes_1 k \otimes_3 f'$$
$$- 1 \otimes_3 1 \otimes_1 1 \otimes_2 1 - 1 \otimes_3 i \otimes_1 i \otimes_2 f'$$
$$- 1 \otimes_3 j \otimes_1 j \otimes_2 1 - 1 \otimes_3 k \otimes_1 k \otimes_2 f')$$
$$= \frac{1}{6}(1 \otimes_1 1 \otimes_2 1 \otimes_3 f' - 1 \otimes_2 1 \otimes_1 1 \otimes_3 f'$$
$$+ 1 \otimes_2 1 \otimes_3 1 \otimes_1 f' - 1 \otimes_3 1 \otimes_2 1 \otimes_1 f'$$
$$+ 1 \otimes_3 1 \otimes_1 1 \otimes_2 f' - 1 \otimes_1 1 \otimes_3 1 \otimes_2 f')$$

**February 24.** Sudbery argued differently. He determined forms according to their values for given arguments. I can to look what I will get.

I will start from differential form $Dq$. Sudbery considers the form

$$v = dx^0 \wedge dx^1 \wedge dx^2 \wedge dx^3$$

## 2.4. Cauchy-Riemann-Fueter Equation

In my notation
$$v = [1 \otimes_1 1 \otimes_2 1 \otimes_3 1 \otimes_4 1]$$
It is interesting to observe how familiar notation takes another form. There is separate question how it can help.

I am very confused by equality [21]-(2.27). It has form

(2.4.13) $\quad <h_1, Dq \circ (h_2, h_3, h_4)> = v \circ (h_1, h_2, h_3, h_4)$

Right part is skew symmetric with respect to all variables, however I cannot tell that
$$<h_1, Dq \circ (h_2, h_3, h_4)> = -<h_2, Dq \circ (h_1, h_3, h_4)>$$
However, according to the definition (2.4.10), differential form $Dq$ is not like an expression which is skew symmetric in all variables.

**February 25.** I realized that I do not move ahead and turned for help to the professor Sudbery. He drew my attention that

$$\begin{aligned} Dq \circ (a, b, c) - Dq \circ (bac) &= \frac{1}{2}(ca^*b - ba^*c + cb^*a - ab^*c) \\ &= \frac{1}{2}(ca^*b + cb^*a - ba^*c - ab^*c) \\ &= \frac{1}{2}(c(a^*b + b^*a) - (ba^* + ab^*)c) = 0 \end{aligned}$$

because
$$a^*b + b^*a = a^*b + (a^*b^{**})^* = \operatorname{Re}(a^*b) = \operatorname{Re}(ba^*) = ba^* + (b^{**}a^*)^* = ba^* + ab^*$$

Today in the search for more information on this topic, I found the paper [22]. I found more details in this paper. In the paper [22], Sudbery uses the equality

(2.4.14) $\quad <a, b> = \frac{1}{2}(a^*b + b^*a)$

The equality

$$< a, Dq \circ (b, c, d) > + < b, Dq \circ (a, c, d) >$$
$$= \frac{1}{2}(a^* Dq \circ (b, c, d) > + (Dq \circ (b, c, d))^* a$$
$$+ b^* Dq \circ (a, c, d) > + (Dq \circ (a, c, d))^* b)$$
$$= \frac{1}{4}(a^*(db^*c - cb^*d) + (db^*c - cb^*d)^* a$$
$$+ b^*(da^*c - ca^*d) + (da^*c - ca^*d)^* b)$$
(2.4.15)
$$= \frac{1}{4}(a^* db^* c - a^* cb^* d + c^* bd^* a - d^* bc^* a$$
$$+ b^* da^* c - b^* ca^* d + c^* ad^* b - d^* ac^* b)$$
$$= \frac{1}{4}(a^* db^* c + c^* bd^* a - a^* cb^* d - d^* bc^* a$$
$$+ b^* da^* c + c^* ad^* b - b^* ca^* d - d^* ac^* b)$$
$$= \frac{1}{4}(a^* db^* c + (a^* db^* c)^* - a^* cb^* d - (a^* cb^* d)^*$$
$$+ b^* da^* c + (b^* da^* c)^* - b^* ca^* d - (b^* ca^* d)^*)$$
$$= \frac{1}{4}(\operatorname{Re}(a^* db^* c) - \operatorname{Re}(a^* cb^* d) + \operatorname{Re}(b^* da^* c) - \operatorname{Re}(b^* ca^* d)) = 0$$

follows from the equalities (2.4.10), (2.4.14). So, in left and right parts of the equality (2.4.13) we have skew symmetric maps.

**February 26.** Picture becomes clearer. As I understand, polylinear maps $dq$, $dq \wedge dq$, $Dq$, $v$ are intended as volume elements in the evaluation of the integral. The main my problem is that integrand is differential form. I must keep the correct correspondence of the parameters.

But before I will continue, I need to finish the study of Fueter equation.

Now I want to make sure that I have wrote properly the expression for $Dq$.

(2.4.16)
$$Dq \circ (a, b, c) = \frac{1}{4}(1 \otimes_2 1 \otimes_1 1 \otimes_3 1 + 1 \otimes_2 i \otimes_1 i \otimes_3 1$$
$$+ 1 \otimes_2 j \otimes_1 j \otimes_3 1 + 1 \otimes_2 k \otimes_1 k \otimes_3 1$$
$$- 1 \otimes_3 1 \otimes_1 1 \otimes_2 1 - 1 \otimes_3 i \otimes_1 i \otimes_2 1$$
$$- 1 \otimes_3 j \otimes_1 j \otimes_2 1 - 1 \otimes_3 k \otimes_1 k \otimes_2 1) \circ (a, b, c)$$
$$= \frac{1}{4}(bac + biaic + bjajc + bkakc$$
$$- cab - ciaib - cjajb - ckakb)$$
$$= \frac{1}{4}(b(a + iai + jaj + kak)c$$
$$- c(a + iai + jaj + kak)b)$$

(2.4.17)
$$\begin{aligned}
a + iai + jaj + kak &= a^0 + a^1i + a^2j + a^3k + i(a^0 + a^1i + a^2j + a^3k)i \\
&\quad + j(a^0 + a^1i + a^2j + a^3k)j \\
&\quad + k(a^0 + a^1i + a^2j + a^3k)k \\
&= a^0 + a^1i + a^2j + a^3k - a^0 - a^1i + a^2j + a^3k \\
&\quad - a^0 + a^1i - a^2j + a^3k - a^0 + a^1i + a^2j - a^3k \\
&= -2a^0 + 2a^1i + 2a^2j + 2a^3k
\end{aligned}$$

The equality

(2.4.18)
$$\begin{aligned}
Dq \circ (a,b,c) &= \frac{1}{2}((c^0 + c^1i + c^2j + c^3k)(a^0 - a^1i - a^2j - a^3k)b \\
&\quad - b(a^0 - a^1i - a^2j - a^3k)(c^0 + c^1i + c^2j + c^3k)) \\
&= \frac{1}{2}(c^0a^0 - c^0a^1i - c^0a^2j - c^0a^3k \\
&\quad + c^1a^0i - c^1a^1ii - c^1a^2ij - c^1a^3ik \\
&\quad + c^2a^0j - c^2a^1ji - c^2a^2jj - c^2a^3jk \\
&\quad + c^3a^0k - c^3a^1ki - c^3a^2kj - c^3a^3kk)b \\
&\quad - b(a^0c^0 + a^0c^1i + a^0c^2j + a^0c^3k \\
&\quad - a^1c^0i - a^1c^1ii - a^1c^2ij - a^1c^3ik \\
&\quad - a^2c^0j - a^2c^1ji - a^2c^2jj - a^2c^3jk \\
&\quad - a^3c^0k - a^3c^1ki - a^3c^2kj - a^3c^3kk)
\end{aligned}$$

follows from equalities (2.4.16), (2.4.17). We reduce the equality (2.4.18)

$$\begin{aligned}
Dq \circ (a,b,c) &= \frac{1}{2}(c^0a^0 - c^0a^1i - c^0a^2j - c^0a^3k + c^1a^0i + c^1a^1 - c^1a^2k + c^1a^3j \\
&\quad + c^2a^0j + c^2a^1k + c^2a^2 - c^2a^3i + c^3a^0k - c^3a^1j + c^3a^2i + c^3a^3)b \\
&\quad - b(a^0c^0 + a^0c^1i + a^0c^2j + a^0c^3k - a^1c^0i + a^1c^1 - a^1c^2k + a^1c^3j \\
&\quad - a^2c^0j + a^2c^1k + a^2c^2 - a^2c^3i - a^3c^0k - a^3c^1j + a^3c^2i + a^3c^3)
\end{aligned}$$

$$\begin{aligned}
Dq \circ (a,b,c) &= \frac{1}{2}(c^0a^0 + c^1a^1 + c^2a^2 + c^3a^3 + (-c^0a^1 + c^1a^0 - c^2a^3 + c^3a^2)i \\
&\quad + (-c^0a^2 + c^1a^3 + c^2a^0 - c^3a^1)j \\
&\quad + (c^2a^1 - c^0a^3 - c^1a^2 + c^3a^0)k)(b^0 + b^1i + b^2j + b^3k) \\
&\quad - (b^0 + b^1i + b^2j + b^3k)(a^0c^0 + a^1c^1 + a^2c^2 + a^3c^3 \\
&\quad + (a^0c^1 - a^1c^0 - a^2c^3 + a^3c^2)i \\
&\quad + (a^0c^2 + a^1c^3 - a^2c^0 - a^3c^1)j + (a^0c^3 - a^1c^2 + a^2c^1 - a^3c^0)k)
\end{aligned}$$

$$\begin{aligned}
2Dq \circ (a,b,c) = &\ (c^0 a^0 + c^1 a^1 + c^2 a^2 + c^3 a^3) b^0_{-1-} \\
&+ (-c^0 a^1_{-5-} + c^1 a^0_{-6-} - c^2 a^3_{=2=} + c^3 a^2_{=1=}) b^0 i \\
&+ (-c^0 a^2_{-8-} + c^1 a^3_{=3=} + c^2 a^0_{-7-} - c^3 a^1_{=4=}) b^0 j \\
&+ (c^2 a^1_{=5=} - c^0 a^3_{-9-} - c^1 a^2_{=6=} + c^3 a^0_{-10-}) b^0 k \\
&+ (c^0 a^0 + c^1 a^1 + c^2 a^2 + c^3 a^3) b^1 i_{-2-} \\
&+ (-c^0 a^1_{-11-} + c^1 a^0_{-12-} - c^2 a^3_{=7=} + c^3 a^2_{=8=}) b^1 ii \\
&+ (-c^0 a^2_{=10=} + c^1 a^3_{-13-} + c^2 a^0_{=9=} - c^3 a^1_{-14-}) b^1 ji \\
&+ (c^2 a^1_{-15-} - c^0 a^3_{=12=} - c^1 a^2_{-16-} + c^3 a^0_{=11=}) b^1 ki \\
&+ (c^0 a^0 + c^1 a^1 + c^2 a^2 + c^3 a^3) b^2 j_{-3-} \\
&+ (-c^0 a^1_{=13=} + c^1 a^0_{=14=} - c^2 a^3_{-17-} + c^3 a^2_{-18-}) b^2 ij \\
&+ (-c^0 a^2_{-20-} + c^1 a^3_{=15=} + c^2 a^0_{-19-} - c^3 a^1_{=16=}) b^2 jj \\
&+ (c^2 a^1_{-21-} - c^0 a^3_{=18=} - c^1 a^2_{-22-} + c^3 a^0_{=17=}) b^2 kj \\
&+ (c^0 a^0 + c^1 a^1 + c^2 a^2 + c^3 a^3) b^3 k_{-4-} \\
&+ (-c^0 a^1_{=20=} + c^1 a^0_{=19=} - c^2 a^3_{-24-} + c^3 a^2_{-23-}) b^3 ik \\
&+ (-c^0 a^2_{=21=} + c^1 a^3_{-25-} + c^2 a^0_{=22=} - c^3 a^1_{-26-}) b^3 jk \\
&+ (c^2 a^1_{=23=} - c^0 a^3_{-27-} - c^1 a^2_{=24=} + c^3 a^0_{-28-}) b^3 kk \\
&- b^0 (a^0 c^0 + a^1 c^1 + a^2 c^2 + a^3 c^3)_{-1-} \\
&- (a^0 c^1_{-6-} - a^1 c^0_{-5-} - a^2 c^3_{=1=} + a^3 c^2_{=2=}) b^0 i \\
&- (a^0 c^2_{-7-} + a^1 c^3_{=4=} - a^2 c^0_{-8-} - a^3 c^1_{=3=}) b^0 j \\
&- (a^0 c^3_{-10-} - a^1 c^2_{=5=} + a^2 c^1_{=6=} - a^3 c^0_{-9-}) b^0 k \\
&- (a^0 c^0 + a^1 c^1 + a^2 c^2 + a^3 c^3) b^1 i_{-2-} \\
&- (a^0 c^1_{-12-} - a^1 c^0_{-11-} - a^2 c^3_{=8=} + a^3 c^2_{=7=}) b^1 ii \\
&- (a^0 c^2_{=9=} + a^1 c^3_{-14-} - a^2 c^0_{=10=} - a^3 c^1_{-13-}) b^1 ij \\
&- (a^0 c^3_{=11=} - a^1 c^2_{-15-} + a^2 c^1_{-16-} - a^3 c^0_{=12=}) b^1 ik \\
&- (a^0 c^0 + a^1 c^1 + a^2 c^2 + a^3 c^3) b^2 j_{-3-} \\
&- (a^0 c^1_{=14=} - a^1 c^0_{=13=} - a^2 c^3_{-18-} + a^3 c^2_{-17-}) b^2 ji \\
&- (a^0 c^2_{-19-} + a^1 c^3_{=16=} - a^2 c^0_{-20-} - a^3 c^1_{=15=}) b^2 jj \\
&- (a^0 c^3_{=17=} - a^1 c^2_{-21-} + a^2 c^1_{-22-} - a^3 c^0_{=18=}) b^2 jk \\
&- (a^0 c^0 + a^1 c^1 + a^2 c^2 + a^3 c^3) b^3 k_{-4-} \\
&- (a^0 c^1_{=19=} - a^1 c^0_{=20=} - a^2 c^3_{-23-} + a^3 c^2_{-24-}) b^3 ki \\
&- (a^0 c^2_{=22=} + a^1 c^3_{-26-} - a^2 c^0_{=21=} - a^3 c^1_{-25-}) b^3 kj \\
&- (a^0 c^3_{-28-} - a^1 c^2_{=23=} + a^2 c^1_{=24=} - a^3 c^0_{-27-}) b^3 kk
\end{aligned}$$

## 2.4. Cauchy-Riemann-Fueter Equation

$$Dq \circ (a,b,c) = (-c^2 a^3 + c^3 a^2)b^0 i + (+c^1 a^3 - c^3 a^1)b^0 j + (c^2 a^1 - c^1 a^2)b^0 k$$
$$- (-c^2 a^3 + c^3 a^2)b^1 - (-c^0 a^2 + c^2 a^0)b^1 k + (-c^0 a^3 + c^3 a^0)b^1 j$$
$$+ (-c^0 a^1 + c^1 a^0)b^2 k - (+c^1 a^3 - c^3 a^1)b^2 - (-c^0 a^3 + c^3 a^0)b^2 i$$
$$- (-c^0 a^1 + c^1 a^0)b^3 j + (-c^0 a^2 + c^2 a^0)b^3 i - (c^2 a^1 - c^1 a^2)b^3$$

**February 28.**

$$Dq \circ (a,b,c) = \begin{vmatrix} a^2 & a^3 \\ c^2 & c^3 \end{vmatrix} b^0 i - \begin{vmatrix} a^1 & a^3 \\ c^1 & c^3 \end{vmatrix} b^0 j + \begin{vmatrix} a^1 & a^2 \\ c^1 & c^2 \end{vmatrix} b^0 k - \begin{vmatrix} a^2 & a^3 \\ c^2 & c^3 \end{vmatrix} b^1$$
$$- \begin{vmatrix} a^0 & a^2 \\ c^0 & c^2 \end{vmatrix} b^1 k + \begin{vmatrix} a^0 & a^3 \\ c^0 & c^3 \end{vmatrix} b^1 j + \begin{vmatrix} a^0 & a^1 \\ c^0 & c^1 \end{vmatrix} b^2 k + \begin{vmatrix} a^1 & a^3 \\ c^1 & c^3 \end{vmatrix} b^2$$
$$- \begin{vmatrix} a^0 & a^3 \\ c^0 & c^3 \end{vmatrix} b^2 i - \begin{vmatrix} a^0 & a^1 \\ c^0 & c^1 \end{vmatrix} b^3 j + \begin{vmatrix} a^0 & a^2 \\ c^0 & c^2 \end{vmatrix} b^3 i - \begin{vmatrix} a^1 & a^2 \\ c^1 & c^2 \end{vmatrix} b^3$$

$$Dq \circ (a,b,c) = - \begin{vmatrix} a^2 & a^3 \\ c^2 & c^3 \end{vmatrix} b^1 + \begin{vmatrix} a^1 & a^3 \\ c^1 & c^3 \end{vmatrix} b^2 - \begin{vmatrix} a^1 & a^2 \\ c^1 & c^2 \end{vmatrix} b^3$$
$$+ \begin{vmatrix} a^2 & a^3 \\ c^2 & c^3 \end{vmatrix} b^0 i - \begin{vmatrix} a^0 & a^3 \\ c^0 & c^3 \end{vmatrix} b^2 i + \begin{vmatrix} a^0 & a^2 \\ c^0 & c^2 \end{vmatrix} b^3 i$$
$$- \begin{vmatrix} a^1 & a^3 \\ c^1 & c^3 \end{vmatrix} b^0 j + \begin{vmatrix} a^0 & a^3 \\ c^0 & c^3 \end{vmatrix} b^1 j - \begin{vmatrix} a^0 & a^1 \\ c^0 & c^1 \end{vmatrix} b^3 j$$
$$+ \begin{vmatrix} a^1 & a^2 \\ c^1 & c^2 \end{vmatrix} b^0 k - \begin{vmatrix} a^0 & a^2 \\ c^0 & c^2 \end{vmatrix} b^1 k + \begin{vmatrix} a^0 & a^1 \\ c^0 & c^1 \end{vmatrix} b^2 k$$

$$Dq \circ (a,b,c) = \begin{vmatrix} a^1 & a^2 & a^3 \\ b^1 & b^2 & b^3 \\ c^1 & c^2 & c^3 \end{vmatrix} - \begin{vmatrix} a^0 & a^2 & a^3 \\ b^0 & b^2 & b^3 \\ c^0 & c^2 & c^3 \end{vmatrix} i$$
$$+ \begin{vmatrix} a^0 & a^1 & a^3 \\ b^0 & b^1 & b^3 \\ c^0 & c^1 & c^3 \end{vmatrix} j - \begin{vmatrix} a^0 & a^1 & a^2 \\ b^0 & b^1 & b^2 \\ c^0 & c^1 & c^2 \end{vmatrix} k$$
$$= a^1 \wedge b^2 \wedge c^3 - (a^0 \wedge b^2 \wedge c^3)i + (a^0 \wedge b^1 \wedge c^3)j - (a^0 \wedge b^1 \wedge c^2)k$$

(2.4.19)
$$Dq \circ (a,b,c) = \begin{vmatrix} 1 & i & j & k \\ a^0 & a^1 & a^2 & a^3 \\ b^0 & b^1 & b^2 & b^3 \\ c^0 & c^1 & c^2 & c^3 \end{vmatrix}$$

I was expected the equality (2.4.19).

(2.4.20) $$Dq \circ (a,b,c) = v \circ (h,a,b,c) \quad h = 1 + i + j + k$$

It is interesting that this vector $h$ is also present in Cauchy-Riemann-Fueter equation. Similarly, the equality

(2.4.21)
$$(dq \wedge dq) \circ (a,b) = \frac{1}{2}\left( \begin{vmatrix} a^2 & a^3 \\ b^2 & b^3 \end{vmatrix} i - \begin{vmatrix} a^1 & a^3 \\ b^1 & b^3 \end{vmatrix} j + \begin{vmatrix} a^1 & a^2 \\ b^1 & b^2 \end{vmatrix} k \right)$$
$$= \frac{1}{2}\begin{vmatrix} i & j & k \\ a^1 & a^2 & a^3 \\ b^1 & b^2 & b^3 \end{vmatrix}$$

follows from the equality (2.4.9). I think it is right

(2.4.22)
$$(dq \wedge dq) \circ (a,b) = \begin{vmatrix} a^2 & a^3 \\ b^2 & b^3 \end{vmatrix} i - \begin{vmatrix} a^1 & a^3 \\ b^1 & b^3 \end{vmatrix} j + \begin{vmatrix} a^1 & a^2 \\ b^1 & b^2 \end{vmatrix} k$$
$$= \begin{vmatrix} i & j & k \\ a^1 & a^2 & a^3 \\ b^1 & b^2 & b^3 \end{vmatrix}$$

And what is next? How does $dq \wedge dq \wedge df$ look like from the point of view of matrix? How does $f'$ look like from the point of view of partial derivative?

In Cauchy-Riemann-Fueter equation, Sudbery uses

(2.4.23) $$f'_l = \frac{1}{2}\left( \frac{\partial f}{\partial x^0} - i\frac{\partial f}{\partial x^1} - j\frac{\partial f}{\partial x^2} - k\frac{\partial f}{\partial x^3} \right)$$

instead of $f'$.

We can represent any differential form as determinant. However equality of matrices do not follow from the equality of determinants. The relation between differential form and determinant is true only for commutative ring. This relation is not true for non commutative algebra.

**March 2.** I had an option how to rewrite Cauchy-Riemann-Fueter equation. But derivative $f'_l$ in right part breaks my plans. However, I will finish my calculations. At first, I will use this experience during study of modules over non commutative algebra.

In particular, I was thinking should I represent the volume using quasideterminant. There was time when I doubted it. However at the same time I considered relatively simple case. I considered parallelogram built on two vectors. In what order should I consider the product of these vectors, if product is non commutative.

From a broader perspective, this is a question how represent differential form in module over non commutative algebra.

Why three-dimensional volume is represented by form $Dq$? Why did we select the equality (2.4.1) for generalization Cauchy-Riemann equation?

**March 3.** When I saw the theorem [22]-5 (page 14) for the first time, I was a little confused. The first impression was that this theorem contradicts to the theorem [16]-8.3.4. In fact, there is no contradiction here.

I considered the map $y = ax$. Then I realized that $dqf = a$ and everything is fine here. Evidently there is no function $f$ for which $dqf = ax$, because function $y = x^2$ does not have left or right derivative.

Today it became obvious that I was moving in the wrong direction. I tried to rewrite the existing formula using different language. It works in some cases. But not in this case. But the work which was done will help me in the future. In particular, I realized that the root of problem is in the theory of integration.

Sudbery and Fueter consider Stokes' theorem in quaternion algebra. They consider following equalities

$$d(dq\, f) = dq \wedge dq\, f'_l \tag{2.4.24}$$

$$d(dq \wedge dq\, f) = Dq\, f' \tag{2.4.25}$$

$$d(Dq\, f) = v\, f'_l \tag{2.4.26}$$

We can write these equalities as integrals

$$\int dq\, f = \int dq \wedge dq\, f'_l \tag{2.4.27}$$

$$\int dq \wedge dq\, f = \int Dq\, f' \tag{2.4.28}$$

$$\int Dq\, f = \int v\, f'_l \tag{2.4.29}$$

According to the theorem [22]-5 (page 14), the equality (2.4.24) for Sudbery and Fueter is not interesting. It is not clear why they did not consider the equality (2.4.29).

I suppose to consider all three equalities.

For simplicity, I start with the equality (2.4.24). Therefore, I need to calculate integrals (2.4.27).

I will make a stop here. This is the place where I misinterpreted the calculations in Sudbery paper. There is scalar function in integral (2.4.27). It is right. Sudbery considers left side derivative. If I would confine myself to definite integrals, it would be right. Such integral exists from point of view of integral theory.

However, I consider integrals generated by differential form. In particular, I consider indefinite integral. Therefore, in the equality (2.4.27) in left side there must be not a scalar, but a tensor. I should see similar changes in the remaining equalities.

So the equality (2.4.27) gets form

$$(2.4.30) \qquad \int a \circ dq = ...$$

If $a = 1 \otimes f$, then the equality (2.4.30) gets form (2.4.27). This will preserve succession. Then it will be possible to consider other questions.

**March 4.** I will start with the map $f(x) = x$ and consider the contour

$$\gamma_1(t) = ti$$
$$\gamma_2(t) = i + tj$$
$$\gamma_3(t) = i + j - ti$$
$$\gamma_4(t) = j - tj$$
$$t \in [0, 1]$$

$$(2.4.31) \qquad \int_{\gamma_1} dx\, x = \int_0^1 dt(1 \otimes ti) \circ i = \int_0^1 dt\, ti\, i = -\int_0^1 t\, dt = -\frac{1}{2}t^2\Big|_0^1 = -\frac{1}{2}$$

$$(2.4.32) \qquad \int_{\gamma_2} dx\, x = \int_0^1 dt(1 \otimes (i+tj)) \circ j = \int_0^1 dt\, j(i+tj) = \frac{1}{2}(i+tj)^2\Big|_0^1$$
$$= \frac{1}{2}((j+i)^2 - i) = -\frac{1}{2}(2k+1)$$

$$(2.4.33) \qquad \int_{\gamma_3} dx\, x = \int_0^1 dt(1 \otimes (i+j-ti)) \circ i = -\int_0^1 dt\, i(i+j-ti)$$
$$= \frac{1}{2}(i+j-ti)^2\Big|_0^1 = \frac{1}{2}(j^2 - (i+j)^2) = -\frac{1}{2}i(2j+i)$$
$$= \frac{1}{2}(1-2k)$$

$$(2.4.34) \qquad \int_{\gamma_4} dx\, x = \int_0^1 dt(1 \otimes (j-tj)) \circ (-j) = -\int_0^1 dt\, j(j-tj)$$
$$= \frac{1}{2}(j-tj)^2\Big|_0^1 = -\frac{1}{2}j^2 = \frac{1}{2}$$

$$\int_\gamma dx\, x = \int_{\gamma_1} dx\, x + \int_{\gamma_2} dx\, x + \int_{\gamma_3} dx\, x + \int_{\gamma_4} dx\, x$$

(2.4.35)
$$= -\frac{1}{2} - \frac{1}{2}(2k+1) + \frac{1}{2}(1-2k) + \frac{1}{2}$$
$$= -k$$

Now I can move to the right part. However, I have one remark before. I am working with the equality like

(2.4.36)
$$\int_{\partial S} \omega \circ dx = \int_S d\omega \circ (dx \wedge dx)$$

The equality (2.4.36) is Stokes' theorem. Since $d\omega$ is skew symmetric, then

$$d\omega \circ (dx \wedge dx) = d\omega \circ (dx, dx)$$

Now I consider area

$$S = \{x : x^0 = x^3 = 0, 0 \le x^1 \le 1, 0 \le x^2 \le 1\}$$

and the integral

(2.4.37)
$$\int_S (1 \otimes_1 1 \otimes_2 1 - 1 \otimes_2 1 \otimes_1 1) \circ (dx\, i \wedge dy\, j)$$
$$= \int_0^1 \int_0^1 (1 \otimes_1 1 \otimes_2 1 - 1 \otimes_2 1 \otimes_1 1) \circ (dx\, i, dy\, j)$$

I have impression that this integral is equal to $k$.

**March 5.** The volume is correct up to orientation. But now the error in the sign does not bother me. I will return to this question. It is evident that the equality is true for any dimension.

It is also necessary to pay attention to the fact that after Fueter defined a regular function by the equality (2.4.25), he uses regular map in following equalities.

- If the function $f$ is regular at every point of 4-parallelepiped $C$, then

(2.4.38)
$$\int_{\partial C} Dq\, f = 0$$

  (the theorem [22]-9, page 18)

- If the function $f$ is regular at every point of 4-parallelepiped $C$ and H-number $q_0$ is inner point of 4-parallelepiped $C$, then

(2.4.39)
$$f(q_0) = \frac{1}{2\pi^2} \int_{\partial C} \frac{(q-q_0)^{-1}}{|q-q_0|^2} Dq\, f(q)$$

  (the theorem [22]-10, page 19)

The most interesting for me was the theorem [22]-7, page 16, which states that for non-generate variable 3-parallelepiped $C$

(2.4.40)
$$\lim \int_C Dq = 0 \Rightarrow \lim \left( \int_C Dq \right)^{-1} \left( \int_{\partial C} dq \wedge dq f \right) = 2f'$$

Comparing with the equality [16]-(3.3.2), I can assume that the definition of derivative does not depend on dimension of parallelepiped. Is it right? Is this statement true for any algebra?

**March 6.** Today I decided to read the article [5]. Correspondingly, I will use little different notation. In particular, Deavours uses notation $dV$ instead of $dv$, Deavours uses notation $Dq$ instead of $dQ$. Deavours introduces also gradient operator
$$\Box = \frac{\partial}{\partial x^0} + i\frac{\partial}{\partial x^1} + j\frac{\partial}{\partial x^2} + k\frac{\partial}{\partial x^3}$$
In particular
$$\Box f = 2\overline{\partial}_l f$$
We see notation of Sudbery in right part.

Stokes' theorem has the following form
$$\int_{\partial \sigma} dQ\, f = \int_\sigma \Box f\, dV \tag{2.4.41}$$
(the theorem [5]-2.1). Up to the order of factors, the equation (2.4.41) coincides with equation (2.4.29).

On the page [5]-999, Deavours shows analogy which is important for me. The map $f$ of complex field is regular if
$$\int_C f(z) dz = 0 \tag{2.4.42}$$
for any closed contour $C$. Similarly, the map $f$ of quaternion algebra is left regular if
$$\int_{\partial \sigma} dQ\, f = 0 \tag{2.4.43}$$
for any closed hypersurface $\partial \sigma$. The equation (4) in the theorem [5]-3.1 coincides with Cauchy-Riemann-Fueter equation.

The lemma [5]-3.2. is interesting for me. This lemma states that if $f$ is right regular and $g$ is left regular, then
$$\int_{\partial \sigma} f\, dQ\, g = 0 \tag{2.4.44}$$
I can write the equality (2.4.44) as
$$\int_{\partial \sigma} (f \otimes g)\, dQ = 0 \tag{2.4.45}$$
This definition of left regular map coincides with the theorem [22]-9 on page 18. I can formulate the theorem [22]-8 on page 18 the following way.

THEOREM 2.4.1. *Differentiable map $f$ is regular iff*
$$dQ \wedge \left(\frac{df}{dx} \circ dx\right) = 0 \tag{2.4.46}$$

## 2.4. Cauchy-Riemann-Fueter Equation

It seems that I have everything that I need.

**March 7.** Before I move ahead, I want to make sure that Cauchy-Riemann-Fueter equation follows from the equality (2.4.46).

(2.4.47)
$$Dq \circ (a,b,c)(d\,f') = \left( \begin{vmatrix} a^1 & a^2 & a^3 \\ b^1 & b^2 & b^3 \\ c^1 & c^2 & c^3 \end{vmatrix} - \begin{vmatrix} a^0 & a^2 & a^3 \\ b^0 & b^2 & b^3 \\ c^0 & c^2 & c^3 \end{vmatrix} i \right.$$
$$\left. + \begin{vmatrix} a^0 & a^1 & a^3 \\ b^0 & b^1 & b^3 \\ c^0 & c^1 & c^3 \end{vmatrix} j - \begin{vmatrix} a^0 & a^1 & a^2 \\ b^0 & b^1 & b^2 \\ c^0 & c^1 & c^2 \end{vmatrix} k \right)$$
$$* \left( d^0 \frac{\partial f}{\partial x^0} + d^1 i \frac{\partial f}{\partial x^1} + d^2 j \frac{\partial f}{\partial x^2} + d^3 k \frac{\partial f}{\partial x^3} \right)$$

I think I would better use another representation.

$$Dq \circ (a,b,c)(d\,f') - Dq \circ (a,d,c)(b\,f')$$

(2.4.48)
$$= \left( -\begin{vmatrix} a^2 & a^3 \\ c^2 & c^3 \end{vmatrix} b^1 + \begin{vmatrix} a^1 & a^3 \\ c^1 & c^3 \end{vmatrix} b^2 - \begin{vmatrix} a^1 & a^2 \\ c^1 & c^2 \end{vmatrix} b^3 + \begin{vmatrix} a^2 & a^3 \\ c^2 & c^3 \end{vmatrix} b^0 i \right.$$
$$- \begin{vmatrix} a^0 & a^3 \\ c^0 & c^3 \end{vmatrix} b^2 i + \begin{vmatrix} a^0 & a^2 \\ c^0 & c^2 \end{vmatrix} b^3 i - \begin{vmatrix} a^1 & a^3 \\ c^1 & c^3 \end{vmatrix} b^0 j + \begin{vmatrix} a^0 & a^3 \\ c^0 & c^3 \end{vmatrix} b^1 j$$
$$\left. - \begin{vmatrix} a^0 & a^1 \\ c^0 & c^1 \end{vmatrix} b^3 j + \begin{vmatrix} a^1 & a^2 \\ c^1 & c^2 \end{vmatrix} b^0 k - \begin{vmatrix} a^0 & a^2 \\ c^0 & c^2 \end{vmatrix} b^1 k + \begin{vmatrix} a^0 & a^1 \\ c^0 & c^1 \end{vmatrix} b^2 k \right)$$
$$* \left( d^0 \frac{\partial f}{\partial x^0} + d^1 i \frac{\partial f}{\partial x^1} + d^2 j \frac{\partial f}{\partial x^2} + d^3 k \frac{\partial f}{\partial x^3} \right)$$
$$- \left( -\begin{vmatrix} a^2 & a^3 \\ c^2 & c^3 \end{vmatrix} d^1 + \begin{vmatrix} a^1 & a^3 \\ c^1 & c^3 \end{vmatrix} d^2 - \begin{vmatrix} a^1 & a^2 \\ c^1 & c^2 \end{vmatrix} d^3 + \begin{vmatrix} a^2 & a^3 \\ c^2 & c^3 \end{vmatrix} d^0 i \right.$$
$$- \begin{vmatrix} a^0 & a^3 \\ c^0 & c^3 \end{vmatrix} d^2 i + \begin{vmatrix} a^0 & a^2 \\ c^0 & c^2 \end{vmatrix} d^3 i - \begin{vmatrix} a^1 & a^3 \\ c^1 & c^3 \end{vmatrix} d^0 j + \begin{vmatrix} a^0 & a^3 \\ c^0 & c^3 \end{vmatrix} d^1 j$$
$$\left. - \begin{vmatrix} a^0 & a^1 \\ c^0 & c^1 \end{vmatrix} d^3 j + \begin{vmatrix} a^1 & a^2 \\ c^1 & c^2 \end{vmatrix} d^0 k - \begin{vmatrix} a^0 & a^2 \\ c^0 & c^2 \end{vmatrix} d^1 k + \begin{vmatrix} a^0 & a^1 \\ c^0 & c^1 \end{vmatrix} d^2 k \right)$$
$$* \left( b^0 \frac{\partial f}{\partial x^0} + b^1 i \frac{\partial f}{\partial x^1} + b^2 j \frac{\partial f}{\partial x^2} + b^3 k \frac{\partial f}{\partial x^3} \right)$$

I write coefficient for $\dfrac{\partial f}{\partial x^0}$

$$
\begin{aligned}
(2.4.49) \quad &\begin{vmatrix} a^2 & a^3 \\ c^2 & c^3 \end{vmatrix}\begin{vmatrix} b^0 & b^1 \\ d^0 & d^1 \end{vmatrix} - \begin{vmatrix} a^1 & a^3 \\ c^1 & c^3 \end{vmatrix}\begin{vmatrix} b^0 & b^2 \\ d^0 & d^2 \end{vmatrix} + \begin{vmatrix} a^1 & a^2 \\ c^1 & c^2 \end{vmatrix}\begin{vmatrix} b^0 & b^3 \\ d^0 & d^3 \end{vmatrix} \\
&+ \begin{vmatrix} a^0 & a^3 \\ c^0 & c^3 \end{vmatrix}\begin{vmatrix} b^0 & b^2 \\ d^0 & d^2 \end{vmatrix} i - \begin{vmatrix} a^0 & a^2 \\ c^0 & c^2 \end{vmatrix}\begin{vmatrix} b^0 & b^3 \\ d^0 & d^3 \end{vmatrix} i + \begin{vmatrix} a^0 & a^3 \\ c^0 & c^3 \end{vmatrix}\begin{vmatrix} b^0 & b^1 \\ d^0 & d^1 \end{vmatrix} j \\
&+ \begin{vmatrix} a^0 & a^1 \\ c^0 & c^1 \end{vmatrix}\begin{vmatrix} b^0 & b^3 \\ d^0 & d^3 \end{vmatrix} j - \begin{vmatrix} a^0 & a^2 \\ c^0 & c^2 \end{vmatrix}\begin{vmatrix} b^0 & b^1 \\ d^0 & d^1 \end{vmatrix} k + \begin{vmatrix} a^0 & a^1 \\ c^0 & c^1 \end{vmatrix}\begin{vmatrix} b^0 & b^2 \\ d^0 & d^2 \end{vmatrix} k
\end{aligned}
$$

I have to do a permutation for all $H$-numbers to see my expectation. But this is not necessary. I know now what I will see.

The equation I am looking for has the form

$$
(2.4.50) \quad \frac{d_{s\cdot 0}f}{dx}\frac{d_{s\cdot 1}f}{dx} + i\frac{d_{s\cdot 0}f}{dx}i\frac{d_{s\cdot 1}f}{dx} + j\frac{d_{s\cdot 0}f}{dx}j\frac{d_{s\cdot 1}f}{dx} + k\frac{d_{s\cdot 0}f}{dx}k\frac{d_{s\cdot 1}f}{dx}
$$

But it is too early to rejoice. I can write the equation (2.4.50) as

$$
(2.4.51) \quad \frac{df}{dx}\circ 1 + i\frac{df}{dx}\circ i + j\frac{df}{dx}\circ j + k\frac{df}{dx}\circ k = 0
$$

But I saw this equation a year ago.

At least, I understand how this equation appeared. I do not think that it makes sense to return to this topic in the future. The equality (2.4.45) may be interesting choice. But I need to understand if the residue theorem is true for the map $f\otimes g$.

## 2.5. $D$-Algebra

**March 16.** When we study algebra, we gradually give up some properties of product. We study algebras in the same order. Initially we learn real and complex field. Then we learn quaternion algebra which is not commutative. And finally we learn octonion algebra which is not commutative and is not associative.

So the statement that associativity follows from commutativity appears. But can we prove or refute this statement?

Suddenly I got the answer from the book [6]. According to the definition, Jordan algebra is the algebra with commutative, but nonassociative product such that
$$x^2 \bullet (y \bullet x) = (x^2 \bullet y) \bullet x$$
I do not suppose to consider now example of Jordan algebra. But I want to consider example of commutative nonassociative algebra. I will use the definition of product according to Jordan

$$(2.5.1) \quad a \bullet b = \frac{1}{2}(ab + ba)$$

## 2.5. D-Algebra

Consider the set $S(3)$ of matrices of order 3 and choose matrices

$$a = \begin{pmatrix} 0 & 1 & 0 \\ 1 & 0 & 0 \\ 0 & 0 & 1 \end{pmatrix} \quad b = \begin{pmatrix} 1 & 0 & 0 \\ 0 & 0 & 1 \\ 0 & 1 & 0 \end{pmatrix} \quad c = \begin{pmatrix} 0 & 0 & 1 \\ 1 & 0 & 0 \\ 0 & 1 & 0 \end{pmatrix}$$

Then

$$a \bullet b = \frac{1}{2}\left(\begin{pmatrix} 0 & 0 & 1 \\ 1 & 0 & 0 \\ 0 & 1 & 0 \end{pmatrix} + \begin{pmatrix} 0 & 1 & 0 \\ 0 & 0 & 1 \\ 1 & 0 & 0 \end{pmatrix}\right) = \frac{1}{2}\begin{pmatrix} 0 & 1 & 1 \\ 1 & 0 & 1 \\ 1 & 1 & 0 \end{pmatrix}$$

$$(a \bullet b) \bullet c = \frac{1}{4}\left(\begin{pmatrix} 1 & 1 & 0 \\ 0 & 1 & 1 \\ 1 & 0 & 1 \end{pmatrix} + \begin{pmatrix} 1 & 1 & 0 \\ 1 & 0 & 1 \\ 0 & 1 & 1 \end{pmatrix}\right) = \frac{1}{4}\begin{pmatrix} 2 & 2 & 0 \\ 1 & 1 & 2 \\ 1 & 1 & 2 \end{pmatrix}$$

$$b \bullet c = \frac{1}{2}\left(\begin{pmatrix} 0 & 0 & 1 \\ 0 & 1 & 0 \\ 1 & 0 & 0 \end{pmatrix} + \begin{pmatrix} 0 & 1 & 0 \\ 1 & 0 & 0 \\ 0 & 0 & 1 \end{pmatrix}\right) = \frac{1}{2}\begin{pmatrix} 0 & 1 & 1 \\ 1 & 1 & 0 \\ 1 & 0 & 1 \end{pmatrix}$$

$$a \bullet (b \bullet c) = \frac{1}{4}\left(\begin{pmatrix} 1 & 1 & 0 \\ 1 & 0 & 1 \\ 0 & 1 & 1 \end{pmatrix} + \begin{pmatrix} 1 & 0 & 1 \\ 1 & 1 & 0 \\ 1 & 1 & 0 \end{pmatrix}\right) = \frac{1}{4}\begin{pmatrix} 2 & 1 & 1 \\ 2 & 1 & 1 \\ 1 & 2 & 1 \end{pmatrix}$$

I wanted to see exactly such equalities.

**May 6.** Let *Basise* be basis of vector space $V$ over $D$-algebra $A$. We can represent linear map $f$ of $A$-vector space $V$ using matrix of linear maps of $D$-algebra $A$. If $D$-algebra $A$ is associative, then we represent product of maps $f$ and $g$ using product of their matrices.

However, what we can do if $D$-algebra $A$ is not associative? Unexpected solution follows from the definition 2.5.1.

DEFINITION 2.5.1. *The* **associator**

(2.5.2) $$(a, b, c) = (ab)c - a(bc)$$

*measures associativity in $D$-algebra $A$. $D$-algebra $A$ is called* **associative**, *if*

$$(a, b, c) = 0$$

□

THEOREM 2.5.2. *Let $\overline{\overline{e}}$ be basis of D-algebra $A$. Then*

(2.5.3) $$(a,b,c) = A_{ijl}^k a^i b^j c^l e_k$$

*where* **coordinates of associator** *are defined by the equality*

(2.5.4) $$A_{ijl}^k = C_{pl}^k C_{ij}^p - C_{ip}^k C_{jl}^p$$

PROOF. The equality

(2.5.5) $$\begin{aligned}(a,b,c) &= (C_{pl}^k (ab)^p c^l - C_{ip}^k a^i (bc)^p) e_k \\ &= (C_{pl}^k C_{ij}^p - C_{ip}^k C_{jl}^p) a^i b^j c^l e_k\end{aligned}$$

follows from the equality (2.5.2). The equality (2.5.4) follows from the equality (2.5.5). □

The most interesting starts here.

Let the map $f$ have the form

(2.5.6) $$b^j = f_i^j \circ a^i$$

Let the map $g$ have the form

(2.5.7) $$c^k = g_j^k \circ b^j$$

When I write the equality

(2.5.8) $$c^k = g_j^k \circ (f_i^j \circ a^i) = (g_j^k \circ f_i^j) \circ a^i$$

in associative algebra, I consider by inertia coordinates of maps as $A$-numbers. However, coordinates of maps are $A \otimes A$-numbers and the equality (2.5.8) gets the form

(2.5.9) $$\begin{aligned}c^k &= (g_{j\,0s}^k \otimes g_{j\,1s}^k) \circ ((f_{i\,0t}^j \otimes f_{i\,1t}^j) \circ a^i) \\ &= ((g_{j\,0s}^k \circ f_{i\,0t}^j) \otimes (f_{i\,1t}^j \circ g_{j\,1s}^k)) \circ a^i\end{aligned}$$

If $D$-algebra $A$ is non associative, then the equality (2.5.9) gets the form

(2.5.10) $$\begin{aligned}c^k &= (g_{j\,0s}^k \otimes g_{j\,1s}^k) \circ ((f_{i\,0t}^j \otimes f_{i\,1t}^j) \circ a^i) \\ &= (g_{j\,0s}^k \otimes g_{j\,1s}^k) \circ ((f_{i\,0t}^j a^i) f_{i\,1t}^j) \\ &= (g_{j\,0s}^k ((f_{i\,0t}^j a^i) f_{i\,1t}^j)) g_{j\,1s}^k\end{aligned}$$

The equality

(2.5.11) $$\begin{aligned}c^k &= (g_{j\,0s}^k \otimes g_{j\,1s}^k) \circ ((f_{i\,0t}^j \otimes f_{i\,1t}^j) \circ a^i) \\ &= ...\end{aligned}$$

follows from equalities (2.5.3), (2.5.10).

The abundance of indexes makes the expression too complex. I will write the equality without indices because indices do not play an essential role here.

$$
\begin{aligned}
A &= (g_0((f_0 a) f_1)) g_1 \\
&= (g_0(f_0(a f_1))) g_1 + (g_0((f_0, a, f_1))) g_1 \\
&= ((g_0 f_0)(a f_1)) g_1 + ((g_0, f_0, (a f_1))) g_1 + (g_0((f_0, a, f_1))) g_1 \\
&= (g_0 f_0)((a f_1) g_1) \\
&\quad + ((g_0 f_0), (a f_1), g_1) + ((g_0, f_0, (a f_1))) g_1 + (g_0((f_0, a, f_1))) g_1
\end{aligned}
$$
(2.5.12)

The expression looks rather cumbersome; I want to return to this problem later.

## 2.6. Anholonomic Time

**March 17.** I want to return to calculation which followed my first research project in university. Before I start, I recall several definitions.

I introduced reference frame in event space of general relativity as continuous field of orthonormal bases. The set of functions $e^l_{(k)}$ is coordinates of basis $\overline{\overline{e}}$ of reference frame with respect to coordinate basis at the same point of event space

$$e^l_{(k)} e^{(k)}_k = \delta^l_k$$

The definition of anholonomity object

(2.6.1) $$c^{(i)}_{(k)(l)} = e^k_{(k)} e^l_{(l)} \left( \frac{\partial e^{(i)}_k}{\partial x^l} - \frac{\partial e^{(i)}_l}{\partial x^k} \right)$$

(2.6.2) $$c^{(i)}_{kl} = \frac{\partial e^{(i)}_k}{\partial x^l} - \frac{\partial e^{(i)}_l}{\partial x^k}$$

followed from the definition of reference frame. There is relation

(2.6.3) $$a^{(k)} = e^{(k)}_k a^k$$

between coordinates of vector field $a$ with respect to reference frame and the set of coordinate bases.

There is relation

(2.6.4) $$\Gamma^{(i)}_{(k)(p)} = e^i_{(k)} e^p_{(p)} e^{(i)}_j \Gamma^j_{ip} - e^i_{(k)} e^p_{(p)} \frac{\partial e^{(i)}_i}{\partial x^p}$$

between components of connection with respect to reference frame and the set of coordinate bases. From equalities (2.6.3), (2.6.4) it follows that we can formally define anholonomic coordinates $x^{(k)}$ such way that the set of maps $e^{(k)}_l$ is Jacobi matrix of the map of coordinates $x^i$ into coordinates $x^{(i)}$

(2.6.5) $$e^{(k)}_l = \frac{\partial x^{(k)}}{\partial x^l}$$

The equality

$$c^{(i)}_{kl} = 0$$

is condition of integrability of the system of differential equations (2.6.5).

We may consider the reference frame as continuous set of observers whose watches are synchronized. Therefore proper time of observer is anholonomic coordinates $x^{(0)}$.

In the section [8]-7.3, I consider simple experiment which confirms anholonomity of time in central gravitational field. We can change this experiment. Let two space ships started from the same point and move in opposite directions. When these space ships meet in another semisphere, their watches will show different time.

**March 18.** The experiment which I described above seems not very convincing. I want to consider the experiment which immediately uses the synchronization procedure.

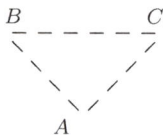

Let $A$, $B$, $C$ be observers. The observer $A$ wants to synchronize watches with the observer $C$ using two procedures. First option is when the observer $A$ sends beam of light immediately to the observer $C$. Second option is when the observer $A$ sends beam of light to the observer $B$, and the observer $B$ sends beam of light to the observer $C$.

The problem of this experiment is that when the observer $A$ sends beam of light to the observer $C$, the observer $A$ must calculate the time of sending of this beam such way that the observer $C$ receives beams of light from observers $A$ and $B$ at the same time. It means that the observer $A$ receives the same value of time of the observer $C$ regardless of the measurement method.

**March 19.** I can change the procedure. Let the observer $A$ periodically send beams of light to the observer $C$ with a predetermined time interval and measure the time interval between received responses. But I have reason to believe that duration of the time interval between received signals will be equal to duration of the time interval between sent signals.

There is one more version of the change. The observer $A$ sends beam of light to the observer $B$. The observer $B$ sends beam of light to the observer $C$. The observer $C$ sends beam of light to the observer $A$. However, when the observer $A$ sends beam of light in the opposite direction, the observer $A$ will get his proper time.

It turns out that the anholonomity of time has nothing to do with synchronization of reference frame. So I will return to the original experiment.

However before I will return to original experiment, I want to make two remarks.

First, what role does synchronization of reference frame play, if this procedure does not help to see real picture? First of all, this procedure allows us to build a set of simultaneous events and to prove that this set depends on choice of reference frame. In general relativity, synchronization of the reference system is

## 2.6. Anholonomic Time

needed for Arnowitt-Deser-Misner formalism. Since at this time I do not consider this formalism, then I can not worry about the procedure of synchronization.

The second remark concerns communication between observers $A$, $B$, $C$. Right now I refreshed in my memory how I determined the speed of light in general relativity. I can use the same procedure to find clock readings when a beam of light will pass through the contour $ABCA$.

The procedure begins with the fact that the observer $A$ exchanges beams of light with observers $B$ and $C$ to determine distance and time intervals $t_{AB}$, $t_{AC}$. Then the observer $A$ exchanges beams of light with the observer $C$ to determine distance and time interval $t_{BC}$. Then the observer $A$ sends beam of light to the observer $B$. The observer $B$ sends beam of light to the observer $C$. The observer $C$ sends beam of light to the observer $A$. Now the observer $A$ can compare the interval $\Delta t$ of time passed from the moment when the observer $A$ sent beam of light to the observer $B$ and the value

$$\Delta t_{AB} + \Delta t_{BC} + \Delta t_{AC}$$

**March 20.** Now I want to estimate anholonomity object in central field of gravity. Let

$$x^0 = t \quad x^1 = r \quad x^2 = \phi \quad x^3 = \theta$$

(2.6.6) $$ds^2 = \frac{x^1 - r_g}{x^1} c^2 d(x^0)^2 - \frac{x^1}{x^1 - r_g} d(x^1)^2 - (x^1)^2 d(x^2)^2 - (x^1)^2 \sin^2 x^2 d(x^3)^2$$

The anholonomic basis has form

$$e_0^{(0)} = c\sqrt{\frac{x^1 - r_g}{x^1}}$$

$$e_1^{(1)} = \sqrt{\frac{x^1}{x^1 - r_g}}$$

$$e_2^{(2)} = x^1$$

$$e_3^{(3)} = x^1 \sin x^2$$

Anholonomity object has form

$$c_{01}^{(0)} = c\frac{\sqrt{r_g}}{x^1}$$

$$c_{21}^{(2)} = 1$$

$$c_{31}^{(3)} = \sin x^2$$

$$c_{32}^{(3)} = x^1 \cos x^2$$

In other words, experiment, which I calculated in the section [8]-7.3 has nothing to do with non-holonomyty.

In order for non-holonomy to manifest itself in the measurement of time, the coordinate $x^1$ must change.

The second variant of the experiment is also incorrect. I do not need to send any beams of light. The experiment is much simpler. I must describe a continuous set of observers, I must to find their proper time and to prove that this function is not continuous.

Let every observer have coordinates
$$x^0 = t \quad x^1 = \text{const} \quad x^2 = \text{const} \quad x^3 = const$$

The proper time $t_0$ of observer at the time $t$ is defined by the equality
$$t_0 = \sqrt{\frac{x^1}{x^1 - r_g}} t$$

The task appeared not easy. By definition

(2.6.7) $$\Delta t_0 = \int dx^{(0)} = \int e_0^{(0)} dx^0$$

This is all, I finally realized. A movement in the radial direction is enough. The observer threw the ball in radial direction; the ball reflected from the wall and returned back to the observer. The observer compared the reading of his watch and of the watch on the ball.

Now we determine parameters of experiment. Let $R_1$ be initial radius, $R_2$ be final radius, $v$ be speed of ball.

Let

(2.6.8) $$x^0 = t \quad x^1 = R_1 + vt \quad 0 \le t \le t_1 \quad R_1 + vt_1 = R_2$$

be initial movement of the ball. The equality

(2.6.9) $$t_1 = \frac{R_2 - R_1}{v}$$

follows from (2.6.8). The equality

(2.6.10) $$\Delta_1 x^{(0)} = \int_0^{t_1} \sqrt{\frac{x^1 - r_g}{x^1}} dx^0 = \int_0^{t_1} \sqrt{\frac{R_1 + vx^0 - r_g}{R_1 + vx^0}} dx^0$$

follows from equalities (2.6.7), (2.6.8), (2.6.9).

**March 21.** I need to get indefinite integral (I changed variable $r = R_1 + vx^0$, $dr = v\, dx^0$ )

(2.6.11) $$\int \sqrt{\frac{R_1 + vx^0 - r_g}{R_1 + vx^0}} dx^0 = \frac{1}{v} \int \sqrt{\frac{r - r_g}{r}} dr$$

Let
$$u^2 = \frac{r - r_g}{r}$$

Then

(2.6.12) $$r = \frac{r_g}{1 - u^2}$$

## 2.6. Anholonomic Time

$$(2.6.13) \qquad dr = \frac{2r_g u}{(1-u^2)^2} du$$

The equality

$$(2.6.14) \qquad \int \sqrt{\frac{R_1 + vx^0 - r_g}{R_1 + vx^0}} dx^0 = \frac{2r_g}{v} \int \frac{u^2}{(1-u^2)^2} du$$

follows from equalities (2.6.11), (2.6.12), (2.6.13). Let

$$(2.6.15) \begin{aligned}
\frac{u^2}{(1-u^2)^2} &= \frac{a}{u-1} + \frac{b}{(u-1)^2} + \frac{c}{u+1} + \frac{d}{(u+1)^2} \\
&= \frac{(a+c)u + a - c}{u^2 - 1} + \frac{(b+d)u^2 + 2(b-d)u + b + d}{(u^2-1)^2} \\
&= \frac{(a+c)u(u^2-1)}{(u^2-1)^2} \\
&\quad + \frac{(a-c+b+d)u^2 + 2(b-d)u + b + d - a + c}{(u^2-1)^2}
\end{aligned}$$

The system of equations

$$(2.6.16) \begin{aligned}
a + c &= 0 \\
a - c + b + d &= 1 \\
b - d &= 0 \\
b + d - a + c &= 0
\end{aligned}$$

follows from the equality (2.6.15). The system of equations

$$(2.6.17) \begin{aligned}
c &= -a \\
b &= d \\
2a + 2b &= 1 \\
2b - 2a &= 0
\end{aligned}$$

follows from the system of equations (2.6.16). The equality

$$(2.6.18) \qquad a = \frac{1}{4} \quad b = \frac{1}{4} \quad c = -\frac{1}{4} \quad d = \frac{1}{4}$$

follows from the system of equations (2.6.17). The equality

$$(2.6.19) \begin{aligned}
\int \sqrt{\frac{R_1 + vx^0 - r_g}{R_1 + vx^0}} dx^0 &= \frac{r_g}{2v} \int \frac{1}{u-1} du + \frac{r_g}{2v} \int \frac{1}{(u-1)^2} du \\
&\quad - \frac{r_g}{2v} \int \frac{1}{u+1} du + \frac{r_g}{2v} \int \frac{1}{(u+1)^2} du \\
&= \frac{r_g}{2v} \left( \ln \frac{u-1}{u+1} - \frac{1}{u-1} - \frac{1}{u+1} \right)
\end{aligned}$$

follows from equalities (2.6.14), (2.6.15), (2.6.18).

If $x^0$ changes from 0 to $t_1$, then $r$ changes from $R_1$ to $R_2$ and $u$ changes from

$$u_1 = \sqrt{\frac{R_1 - r_g}{R_1}} = \sqrt{1 - \frac{r_g}{R_1}}$$

to

$$u_2 = \sqrt{\frac{R_2 - r_g}{R_2}} = \sqrt{1 - \frac{r_g}{R_2}}$$

Because time of movement in both directions is the same, then we get

$$(2.6.20) \qquad \Delta_1 x^{(0)} + \Delta_2 x^{(0)} == \frac{r_g}{v}\left(\ln\frac{u-1}{u+1} - \frac{1}{u-1} - \frac{1}{u+1}\right)_{u=u_1}^{u=u_2}$$

It remains to calculate the time, how long the observer waits return of the ball. The equality

$$(2.6.21) \qquad \Delta_3 x^{(0)} = \int_0^{2t_1} \sqrt{\frac{R_1 - r_g}{R_1}} dx^0 = 2t_1\sqrt{\frac{R_1 - r_g}{R_1}}$$

follows from the equality (2.6.7).

**March 22.** I saved text of programs used for calculation in subsections 2.6.1, 2.6.2. I considered three choices.

In first and second choices the observer is on the surface of the Earth ($R_1 = 6.371e8$ santimeters, the distance from the center of the Earth) and sends signal to the Moon ($R_2 = 3.84e10$ santimeters). The observer sends the ball in first choice. Since the ball must overcome the gravitational field of the Earth, its speed is $112e4$ sm/sec. Coordinate time of travel is 67433.75 seconds. Time measured by the observer on the Earth is 67433.75 seconds. Time measured by the ball is 67433.59 seconds.

Although the difference in the clock readings is very significant, the experiment does not look very real. So I considered second choice, where the observer sends beam of light, $v = 2.998e10$ sm/sec. Coordinate time of travel is 2.5192061374 seconds. Time measured by the observer on the Earth is 2.5192061357 seconds. Time measured by the beam of light is 2.5192002762 seconds. The observer can use interference of the original beam of light and the returned beam of light to check the results of the experiment.

I considered third choice of experiment since I remembered about anomalous acceleration of Pioneer 11 ([1], [2]). So I choose the Sun as central body. The observer is on the Earth ($R_1 = 1.496e13$ santimeters, distance between Sun and Earth) and sends beam of light to Jupiter ($R_2 = 8.17e13$ santimeters, distance between sun and Jupiter). Coordinate time of travel is 4452.30153 seconds. Time measured by the observer on the Earth is 4452.30149 seconds. Time measured by the beam of light is 4452.30138 seconds. Similarly, if the observer sends beam of light to Neptune, then coordinate time of travel is 29022.0147 seconds. Time measured by the observer on the Earth is 29022.0144 seconds. Time measured by the beam of light is 29022.015 seconds. There is a big error in the last number and I cannot improve it.

## 2.6.1. File NonHolonom.aspx.

```
*/%>
<%@ Page Language="C#" AutoEventWireup="true"
    CodeFile="NonHolonom.aspx.cs" Inherits="NonHolonom" %>

<!DOCTYPE html>

<html xmlns="http://www.w3.org/1999/xhtml">
<head runat="server">
    <title></title>
</head>
<body>
    <form id="form1" runat="server">
        <div>
            Mass of Earth <%=CentralMass %>
            <br />
            Gravitational Radius of Earth=<%=GraviRadius %>
            <br />
            Speed of travel 1 =<%=SpeedOfTravel1 %>
            <br />
            Time of travel 1 =<%=TimeOfTravel1 %>
            <br />
            Local time 1 of observer =<%=LocalTime11 %>
            <br />
            Local time 1 of moving object=<%=LocalTime12 %>
            <br /><br />
            Speed of travel 2 =<%=SpeedOfTravel2 %>
            <br />
            Time of travel 2 =<%=TimeOfTravel2 %>
            <br />
            Local time 2 of observer =<%=LocalTime21 %>
            <br />
            Local time 2 of moving object=<%=LocalTime22 %>
            <br /><br />
            Mass of Sun <%=CentralMass3 %>
            <br />
            Gravitational Radius of Sun=<%=GraviRadius3 %>
            <br />
            Speed of travel 3 =<%=SpeedOfTravel2 %>
            <br />
            Time of travel 3 =<%=TimeOfTravel3 %>
            <br />
```

```
                Local time 3 of observer =<%=LocalTime31 %>
                <br />
                Local time 3 of moving object=<%=LocalTime32 %>
                <br /><br />
                Speed of travel 4 =<%=SpeedOfTravel2 %>
                <br />
                Time of travel 4 =<%=TimeOfTravel4 %>
                <br />
                Local time 4 of observer =<%=LocalTime41 %>
                <br />
                Local time 4 of moving object=<%=LocalTime42 %>
        </div>
    </form>
</body>
</html>
<%/*
```

`*/ /*`

### 2.6.2. File NonHolonom.aspx.cs.

```
*/
using System;
using System.Collections.Generic;
using System.Linq;
using System.Web;
using System.Web.UI;
using System.Web.UI.WebControls;

public partial class NonHolonom :   System.Web.UI.Page
{
    public const Double G = 6.673e-8;//cm3 / g c2
    public Double CentralMass, CentralMass3;
    protected Double GraviRadius, GraviRadius3;
    public const Double SpeedOfLight = 2.998e10;//cm/c
    public Double R1,R2;//cm
    protected Double U1,U2;
    public double SpeedOfTravel1, SpeedOfTravel2;
    protected Double TimeOfTravel1, TimeOfTravel2,
        TimeOfTravel3, TimeOfTravel4;
    protected Double LocalTime11, LocalTime12;
    protected Double LocalTime21, LocalTime22;
    protected Double LocalTime31, LocalTime32;
    protected Double LocalTime41, LocalTime42;
    protected void Page_Load(object sender, EventArgs e)
```

## 2.6. Anholonomic Time

```
{
    Double A1, A2;
    CentralMass = 5.977e27;//g mass of Earth
    GraviRadius = 2 * G
        * CentralMass / (SpeedOfLight * SpeedOfLight);
    R1 = 6.371e8;//cm radius of Earth
    R2 = 3.84e10;//cm distance to the Moon
    U1 = Math.Sqrt(1 - GraviRadius / R1);
    U2 = Math.Sqrt(1 - GraviRadius / R2);
    SpeedOfTravel1 = 112e4;//second cosmic speed for Earth
    TimeOfTravel1 = 2 * (R2 - R1) / SpeedOfTravel1;
    LocalTime11 = TimeOfTravel1 * U1;
    A1 = Math.Log((1 - U1) / (1 + U1))
        - 1e0 / (U1 + 1e0) - 1e0 / (U1 - 1);
    A2 = Math.Log((1 - U2) / (1 + U2))
        - 1e0 / (U2 + 1e0) - 1e0 / (U2 - 1);
    LocalTime12 = GraviRadius / SpeedOfTravel1 * (A2 - A1);
    SpeedOfTravel2 = SpeedOfLight;
    TimeOfTravel2 = 2 * (R2 - R1) / SpeedOfTravel2;
    LocalTime21 = TimeOfTravel2 * U1;
    LocalTime22 = GraviRadius / SpeedOfTravel2 * (A2 - A1);
    CentralMass3 = 1.989e33;//g mass of Sun
    GraviRadius3 = 2 * G
        * CentralMass3 / (SpeedOfLight * SpeedOfLight);
    R1 = 1.496e13;//cm distance from the Sun to the Earth
    R2 = 8.17e13;//cm distance from the Sun to the Jupiter
    U1 = Math.Sqrt(1 - GraviRadius3 / R1);
    U2 = Math.Sqrt(1 - GraviRadius3 / R2);
    A1 = Math.Log((1 - U1) / (1 + U1))
        - 1e0 / (U1 + 1e0) - 1e0 / (U1 - 1);
    A2 = Math.Log((1 - U2) / (1 + U2))
        - 1e0 / (U2 + 1e0) - 1e0 / (U2 - 1);
    TimeOfTravel3 = 2 * (R2 - R1) / SpeedOfTravel2;
    LocalTime31 = TimeOfTravel3 * U1;
    LocalTime32 = GraviRadius3 / SpeedOfTravel2 * (A2 - A1);
    R2 = 4.5e14;//cm distance from the Sun to the Jupiter
    U2 = Math.Sqrt(1 - GraviRadius3 / R2);
    A2 = Math.Log((1 - U2) / (1 + U2))
        - 1e0 / (U2 + 1e0) - 1e0 / (U2 - 1);
    TimeOfTravel4 = 2 * (R2 - R1) / SpeedOfTravel2;
    LocalTime41 = TimeOfTravel4 * U1;
    LocalTime42 = GraviRadius3 / SpeedOfTravel2 * (A2 - A1);
}
}
```

## 2.7. Linear Map of Octonion Algebra

**May 20.** I finally wrote the system of linear equations which describes expansion of linear map of octonion algebra with respect to the basis of maps of conjugation. As I expected, the system of linear equations splits into 8 identical systems. So, it is enough to solve only one system. Consider, for instance, the system of linear equations

(2.7.1)
$$\begin{cases} f_0^0 = a_0^0 + a_1^0 + a_2^0 + a_3^0 + a_4^0 + a_5^0 + a_6^0 + a_7^0 \\ f_1^1 = a_0^0 - a_1^0 + a_2^0 + a_3^0 + a_4^0 + a_5^0 + a_6^0 + a_7^0 \\ f_2^2 = a_0^0 + a_1^0 - a_2^0 + a_3^0 + a_4^0 + a_5^0 + a_6^0 + a_7^0 \\ f_3^3 = a_0^0 + a_1^0 + a_2^0 - a_3^0 + a_4^0 + a_5^0 + a_6^0 + a_7^0 \\ f_4^4 = a_0^0 + a_1^0 + a_2^0 + a_3^0 - a_4^0 + a_5^0 + a_6^0 + a_7^0 \\ f_5^5 = a_0^0 + a_1^0 + a_2^0 + a_3^0 + a_4^0 - a_5^0 + a_6^0 + a_7^0 \\ f_6^6 = a_0^0 + a_1^0 + a_2^0 + a_3^0 + a_4^0 + a_5^0 - a_6^0 + a_7^0 \\ f_7^7 = a_0^0 + a_1^0 + a_2^0 + a_3^0 + a_4^0 + a_5^0 + a_6^0 - a_7^0 \end{cases}$$

As in the case of quaternions, it is easy to solve this system of linear equations. It is enough to subtract the first equation from other equations. Therefore

(2.7.2)
$$a_i^0 = \frac{1}{2}(f_0^0 - f_i^i) \quad i = 1, ..., 7$$

The equality

(2.7.3)
$$f_0^0 = a_0^0 + \frac{1}{2}(f_0^0 - f_1^1) + \frac{1}{2}(f_0^0 - f_2^2) + \frac{1}{2}(f_0^0 - f_3^3) + \frac{1}{2}(f_0^0 - f_4^4) \\ + \frac{1}{2}(f_0^0 - f_5^5) + \frac{1}{2}(f_0^0 - f_6^6) + \frac{1}{2}(f_0^0 - f_7^7)$$

follows from equalities (2.7.1), (2.7.2). The equality

$$a_0^0 = \frac{1}{2}(-5f_0^0 + f_1^1 + f_2^2 + f_3^3 + f_4^4 + f_5^5 + f_6^6 + f_7^7)$$

follows from the equality (2.7.3).

## 2.8. Structure of Linear Map

**June 13: Basis of $A \otimes A$-module $\mathcal{L}(D; A \to A)$.** I have been examining for several years the equation

(2.8.1)
$$f_l^k = f^{k \cdot ij} F_{k \cdot l}^{\ m} C_{im}^p C_{pj}^k$$

The main question which is interesting for me follows: using this equation, can I find minimal set of maps $F_k$.

## 2.8. Structure of Linear Map

I recall the original statement.
Consider matrix

$$\tag{2.8.2} \mathcal{C} = (\mathcal{C}^{\cdot k}_{m \cdot ij}) = (C_2{}^{p}_{\cdot im} C_2{}^{k}_{\cdot pj})$$

whose rows and columns are indexed by ${}^{\cdot k}_{m}$ and ${}_{\cdot ij}$, respectively.

The answer is evident when the matrix $\mathcal{C}$ is non singular. In general case, the answer is more complicated. In the beginning, I tried to find the answer in the equation (2.8.1). I chose the equation (2.8.1) because standard components uniquely determine a tensor. However the equation (2.8.1) is not linear with respect to standard components and coordinates of maps $F$.

However, the answer is much simpler. I must to consider the equation

$$\tag{2.8.3} F_{k \cdot l}{}^{m} C^{p}_{im} C^{k}_{pj} = 0$$

It is not too hard to get equation (2.8.3).

Let the set of maps $F_k$ be a basis. If we assume that equation (2.8.1) has 2 solutions, namely $f_1^{k \cdot ij}$ and $f_2^{k \cdot ij}$, then, from the equation (2.8.1), it follows that

$$\tag{2.8.4} 0 = (f_2^{k \cdot ij} - f_1^{k \cdot ij}) F_{k \cdot l}{}^{m} C^{p}_{im} C^{k}_{pj}$$

From the equation (2.8.4), it follows that rank of the matrix $(F_{k \cdot l}{}^{m} C^{p}_{im} C^{k}_{pj})$ less than $n^2$. This is not too much.

**June 14.** My reasoning is wrong. Rank of the matrix $(F_{k \cdot l}{}^{m} C^{p}_{im} C^{k}_{pj})$ is not different from the ramk of the matrix $\mathcal{C}$. So it is not very clear how I get unique expansion, if the addition of a new $F_k$ increases order of the system, but does not change rank. I think, it is mind again to consider complex field.

$$\tag{2.8.5} \begin{aligned} i = 0 \quad j = 0 \quad k = 0 \quad l = 0 \quad F^{m}_{0} C^{p}_{0m} C^{0}_{p0} = 0 \\ F^{0}_{0} C^{0}_{00} C^{0}_{00} = 0 \end{aligned}$$

It is clear that this is nonsense.

**June 15.** The map $F$ does not change rank of the matrix $\mathcal{C}$; however the map changes a set of solutions of the equation (2.8.1). So my goal is to find the set of maps $F$ such that they generate the set of all maps $f \in \mathcal{L}(D; A \to A)$. A set of maps $F$ is finite; however, it would be nice to have simple algorith how to find them.

For any given $k$, the system of linear equations (2.8.1) has rank less than $n^2$. Therefore, there is linear dependence between values $f^{k \cdot ij}$; this dependence is called Riemann equations.

**June 16: Maps of Conjugation.** There is one more interesting question. I named maps $I^i$ as maps of conjugation. Is this name correct?

According to the definition, map of conjugation $I$ satisfies to the equality

$$(2.8.6) \qquad I\circ(ab) = (I\circ b)(I\circ a)$$

Let us verify the map

$$I^1\circ a = a^0 - a^1 i + a^2 j + a^3 k$$

in quaternion algebra. We will start from the equality

$$\begin{aligned}ab =\ & (a^0 b^0 - a^1 b^1 - a^2 b^2 - a^3 b^3)\\ & + (a^1 b^0 + a^0 b^1 - a^3 b^2 + a^2 b^3)i\\ & + (a^2 b^0 + a^3 b^1 + a^0 b^2 - a^1 b^3)j\\ & + (a^3 b^0 - a^2 b^1 + a^1 b^2 + a^0 b^3)k\end{aligned}$$

Then

$$\begin{aligned}(I^1 x)(I^1 a) =\ & (x^0 a^0 - x^1 a^1 - x^2 a^2 - x^3 a^3)\\ & + (-x^1 a^0 - x^0 a^1 - x^3 a^2 + x^2 a^3)i\\ & + (x^2 a^0 - x^3 a^1 + x^0 a^2 + x^1 a^3)j\\ & + (x^3 a^0 + x^2 a^1 - x^1 a^2 + x^0 a^3)k\end{aligned}$$

To my surprise, the equality $(2.8.6)$ is true.

**June 19.** It is more correct to call a map $I^1$ antihomomorphism, because this is the common name.

## 2.9. Non Commutative Sum

**June 16.** We never know where we can meet a new idea.

When I was looking through the calculus textbook, I have met a definition familiar from childhood. This is sum of vectors. The definition is extremely simple.

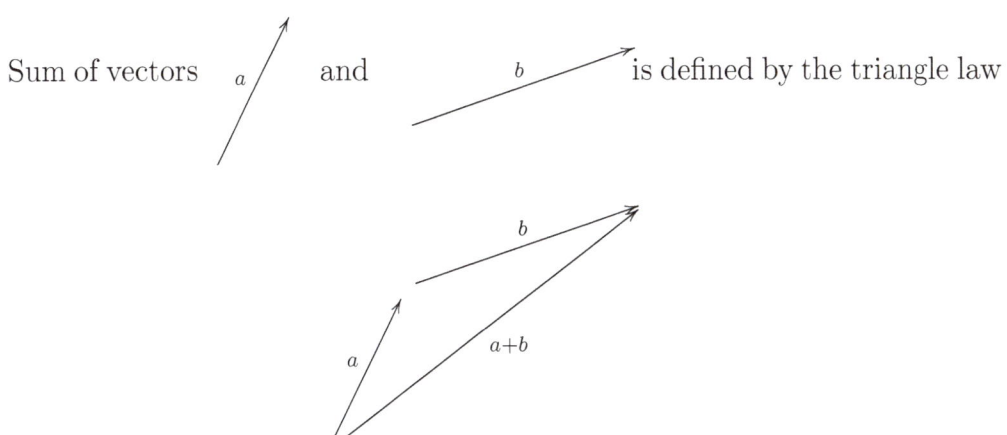

From the parallelogram law it follows that sum is commutative.

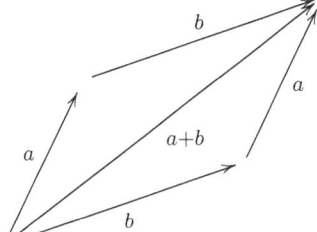

Below we consider a model of affine space in a metric-affine manifold. When we consider connection $\Gamma^k_{ij}$ in Riemann space, we impose a constraint on connection,[2.2] that the torsion

$$T^i_{kl} = \Gamma^i_{lk} - \Gamma^i_{kl} \tag{2.9.1}$$

is 0 (symmetry of connection) and parallel transport does not change scalar product of vectors. If a metric tensor and an arbitrary connection are defined on a differentiable manifold, then this manifold is called **metric-affine manifold**.[2.3] In particular, connection in metric-affine manifold has torsion.

In Riemann space, we use geodesics instead of straight lines. So we can represent the vector $a$ using segment $AB$ of geodesic $L_a$ such that vector $a$ is tangent to geodesic $L_a$ at the point $A$ and length of segment $AB$ equals to length of the vector $a$.

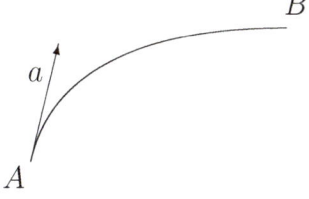

For given vectors $a$ and $b$ in tangent plane at the point $A$, let $\rho$ be the length of the vector $a$ and $\sigma$ be the length of the vector $b$.

---

[2.2] See the definition of affine connection in Riemann space on the page [4]-443.

[2.3] See also the definition [8]-6.1.1.

We transfer the vector $b$ along the segment $AB$ of geodesic from the point $A$ into a point $B$. We mark the result as $b'$. We draw the geodesic $L_{b'}$ through the point $B$ using the vector $b'$ as a tangent vector to $L_{b'}$ in the point $B$. Length of segment $BC$ equals to length of the vector $b$.

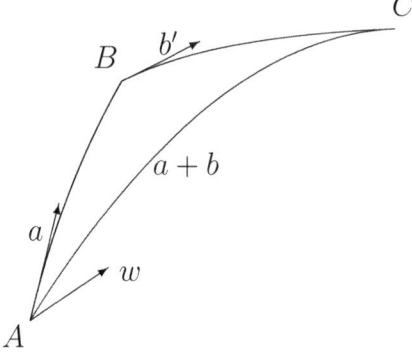

I assume that length of vectors $a$ and $b$ is small. Then there exists unique geodesic $L_c$ from point $A$ to point $C$. I have to prove that the vector $a + b$ is tangent to geodesic $L_c$ and the length of segment $AC$ of geodesic equals to length of vector $a + b$. But at this time I will skip this proof. I will identify segment $AC$ of geodesic and vector $a + b$.

The same way, I draw triangle $ADC$ to find vector $b + a$. Since in Riemann manifold the parallelogram $ABCD$ is closed, then we get the statement that sum of vectors is commutative in Riemann manifold.

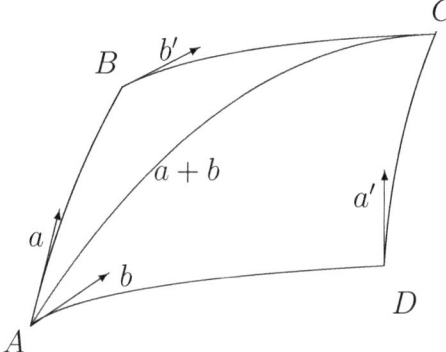

There are two different definitions of a geodesic in the Riemann manifold. One of them relies on the parallel transport. We call an appropriate **line auto parallel**. Another definition depends on the length of trajectory. We call an appropriate **line extreme**. In a metric-affine manifold these lines have different equations.

## 2.9. Non Commutative Sum

To define sum of vectors in metric-affine manifold auto parallel lines are natural choice. However, a parallelogram based on auto parallel lines is not closed in metric-affine manifold (see the theorem [8]-6.2.1). Therefore, sum of vectors in metric-affine manifold is non commutative.

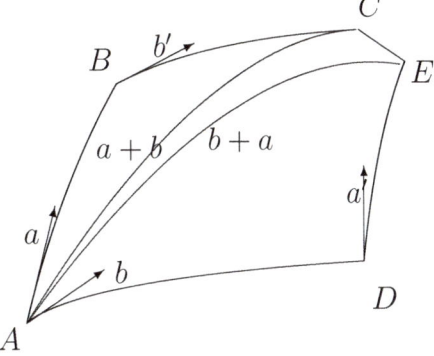

**November 5.** It's time to evaluate the tangent vector of geodesic line $AC$. For given vectors $a$ and $b$ in tangent plane at the point $A$, let $\rho$ be the length of the vector $a$ and $\sigma$ be the length of the vector $b$.

The equation of geodesic line $AB$ has the following form

$$\frac{d^2 x^k}{dt^2} = -\Gamma^k_{ij} \frac{dx^i}{dt} \frac{dx^j}{dt} \tag{2.9.2}$$

with initial condition

$$\left.\frac{dx^i}{dt}\right|_{t=0} = a^i \tag{2.9.3}$$

We also consider parallel translations of the vector $b$ according to the equation

$$\frac{db^k}{dt} = -\Gamma^k_{ij} b^j \frac{dx^i}{dt} \tag{2.9.4}$$

We need the value of the coordinate $x$ and coordinates of the vector $b$ provided

$$t = \rho \quad \rho^2 = g_{ij} a^i a^j$$

Let the value of $\rho$ so small that we can use finite difference method to solve differential equation. Equalities

$$x^k(B) = x^k + a^k \rho - \Gamma^k_{ij} a^i a^j \rho^2 \tag{2.9.5}$$

$$\frac{dx^k}{dt}(B) = a^k - \Gamma^k_{ij} a^i a^j \rho \tag{2.9.6}$$

follow from the differential equation (2.9.2). The equalitiy

$$b^k(B) = b^k - \Gamma^k_{ij} a^i b^j \rho \tag{2.9.7}$$

follows from the differential equation (2.9.4).

The equation of geodesic line $BC$ has the following form

$$\frac{d^2 x^k}{dt^2} = -\Gamma^k_{ij} \frac{dx^i}{dt} \frac{dx^j}{dt} \tag{2.9.8}$$

with initial condition

$$\left.\frac{dx^i}{dt}\right|_{t=0} = b^i(B)$$

We also consider parallel translations of the vector $a$ according to the equation

$$\frac{da^k}{dt} = -\Gamma_{ij}^k a^j \frac{dx^i}{dt} \tag{2.9.9}$$

We need the value of the coordinate $x$ and coordinates of the vector $a$ provided

$$t = \sigma \quad \sigma^2 = g_{ij} b^i b^j$$

Let the value of $\sigma$ so small that we can use finite difference method to solve differential equation. Equalities

$$\begin{aligned}
x^k(C) &= x^k(B) + b^k(B)\sigma - \Gamma_{ij}^k b^i(B) b^j(B) \sigma^2 \\
&= x^k + a^k \rho - \Gamma_{ij}^k a^i a^j \rho^2 + (b^k - \Gamma_{ij}^k a^i b^j \rho)\sigma - \Gamma_{ij}^k b^i b^j \sigma^2 \\
&= x^k + a^k \rho + b^k \sigma - \Gamma_{ij}^k a^i a^j \rho^2 - \Gamma_{ij}^k a^i b^j \rho\sigma - \Gamma_{ij}^k b^i b^j \sigma^2
\end{aligned} \tag{2.9.10}$$

$$\frac{dx^k}{dt}(C) = b^k(B) - \Gamma_{ij}^k b^i(B) b^j(B) \sigma \tag{2.9.11}$$

follow from the differential equation (2.9.8) and from the equalities (2.9.5), (2.9.7).

**November 8.** Let geodesic $AC$ be generated by the vector $c$. Then

$$x^k(C) = x^k(A) + c^k \tau - \Gamma_{ij}^k c^i c^j \tau^2 \tag{2.9.12}$$

Equalities

$$c^k \tau = a^k \rho + b^k \sigma \tag{2.9.13}$$

$$\Gamma_{ij}^k c^i c^j \tau^2 = \Gamma_{ij}^k a^i a^j \rho^2 + \Gamma_{ij}^k a^i b^j \rho\sigma + \Gamma_{ij}^k b^i b^j \sigma^2 \tag{2.9.14}$$

follow from equalities (2.9.10), (2.9.12). Equalities (2.9.13), (2.9.14) look unusual. However, if we assume

$$A^k = a^k \rho \quad B^k = b^k \sigma \quad C^k = c^k \tau$$

then equalities (2.9.13), (2.9.14) will get form

$$C^k = A^k + B^k \tag{2.9.15}$$

$$\Gamma_{ij}^k C^i C^j = \Gamma_{ij}^k A^i A^j + \Gamma_{ij}^k A^i B^j + \Gamma_{ij}^k B^i B^j \tag{2.9.16}$$

Everything is all right. But why, in the equality (2.9.15), do I multiply the coordinates of a vector by the length of this vector? The answer is in format of differential equation (2.9.3). If $t$ is natural parameter of the curve $AB$, then $a^k$ are coordinates of unit vector in the direction of the vector $A^k$. It remains to deal with the equation (2.9.16).

**November 10.** At first sight, the equality (2.9.16) follows from the equality (2.9.15). But it is not true.

$$\begin{aligned}(2.9.17)\quad \Gamma^k_{ij}C^iC^j &= \Gamma^k_{ij}(A^i+B^i)(A^j+B^j)\\ &= \Gamma^k_{ij}(A^iA^j+A^iB^j+B^iA^j+B^iB^j)\end{aligned}$$

From equalities (2.9.16), (2.9.17) it follows that

$$\Gamma^k_{ij}B^iA^j=0$$

Why did it happen?

**November 13.** I see mistake in equalities (2.9.5), (2.9.10). I forgot to write the coefficient $\frac{1}{2}$ in front of $\rho^2$ and $\sigma^2$. So, equalities (2.9.5), (2.9.10) get form

$$(2.9.18)\quad x^k(B) = x^k + a^k\rho - \frac{1}{2}\Gamma^k_{ij}a^ia^j\rho^2$$

$$\begin{aligned}x^k(C) &= x^k(B) + b^k(B)\sigma - \frac{1}{2}\Gamma^k_{ij}b^i(B)b^j(B)\sigma^2\\ (2.9.19)\quad &= x^k + a^k\rho - \frac{1}{2}\Gamma^k_{ij}a^ia^j\rho^2 + (b^k - \Gamma^k_{ij}a^ib^j\rho)\sigma - \frac{1}{2}\Gamma^k_{ij}b^ib^j\sigma^2\\ &= x^k + a^k\rho + b^k\sigma - \frac{1}{2}\Gamma^k_{ij}a^ia^j\rho^2 - \Gamma^k_{ij}a^ib^j\rho\sigma - \frac{1}{2}\Gamma^k_{ij}b^ib^j\sigma^2\end{aligned}$$

The equality (2.9.14) gets form

$$(2.9.20)\quad \Gamma^k_{ij}c^ic^j\tau^2 = \Gamma^k_{ij}a^ia^j\rho^2 + 2\Gamma^k_{ij}a^ib^j\rho\sigma + \Gamma^k_{ij}b^ib^j\sigma^2$$

If connection is symmetric, then the equality (2.9.20) from the equality (2.9.13), because

$$\begin{aligned}(2.9.21)\quad \Gamma^k_{ij}C^iC^j &= \Gamma^k_{ij}(A^i+B^i)(A^j+B^j)\\ &= \Gamma^k_{ij}(A^iA^j+A^iB^j+B^iA^j+B^iB^j)\\ &= \Gamma^k_{ij}(A^iA^j+2A^iB^j+B^iB^j)\end{aligned}$$

That is, the diagonal of the parallelogram is a geodesic.

If connection has torsion, then this geodesic passes by points $C$ and $E$. I will assume that required geodesic are tangent vectors $a+b+d$ and $a+b-d$.

## 2.10. Free Representation

**August 9.** There exist few definitions of free representation.

DEFINITION 2.10.1. *The representation*

$$f: A_1 \xrightarrow{\;*\;} A_2$$

*of the $\Omega_1$-algebra $A_1$ is called* **free**,[2.4] *if the statement*

$$f(a_1)(a_2) = f(b_1)(a_2)$$

*for any $a_2 \in A_2$ implies that $a_1 = b_1$.* □

---
[2.4]See similar definition of free representation of group in [20], page 16.

I want to consider one more definition. This definition is based on the definition of free module. At first I considered the following definition.

DEFINITION 2.10.2. Let $A_1$ be $\Omega_1$-algebra and $X$ be the set. Let $\mathcal{A}(\, A_1 \longrightarrow \mathcal{A}(\Omega_2, X) \,)$ be category. A representation of $\Omega_1$-algebra $A_1$ in $\Omega_2$-algebra $A_2$, $X \subset A_2$, is object of category $\mathcal{A}(\, A_1 \longrightarrow \mathcal{A}(\Omega_2, X) \,)$. Reduced morphism of representation is morphism of category $\mathcal{A}(\, A_1 \longrightarrow \mathcal{A}(\Omega_2, X) \,)$. Universally repelling object of the category $\mathcal{A}(\, A_1 \longrightarrow \mathcal{A}(\Omega_2, X) \,)$ is called **free representation**. □

In such case, from the diagram

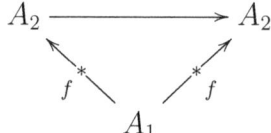

it follows that there exist unique morphism of free representation. It looks like this definition is too restrictive. So I considered another definition.

DEFINITION 2.10.3. Let $A_1$ be $\Omega_1$-algebra and $X$ be the set. Let $\mathcal{A}(\, A_1 \longrightarrow \mathcal{A}(\Omega_2, X) \,)$ be category. A representation of $\Omega_1$-algebra $A_1$ in $\Omega_2$-algebra $A_2$, $X \subset A_2$, is object of category $\mathcal{A}(\, A_1 \longrightarrow \mathcal{A}(\Omega_2, X) \,)$. Reduced morphism of representation is morphism of category $\mathcal{A}(\, A_1 \longrightarrow \mathcal{A}(\Omega_2, X) \,)$. The representation
$$f : A_1 \longrightarrow\!\!\!\ast\!\!\!\longrightarrow A_2$$
is called **free representation**, if for any representation
$$g : A_1 \longrightarrow\!\!\!\ast\!\!\!\longrightarrow B_2$$
reduced morphism of representations
$$h : A_2 \to B_2$$
is defined uniquely up to automorphism of the representation $f$. □

It looks like I have too wide choice. Will the morphism $h$ be an arbitrary morphism. I can use the following diagram to represent this definition

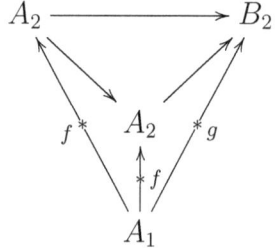

**August 10.** If $\Omega_1 = \emptyset$, then I can identifi representation and universal algebra. So the definition of free universal algebra is interesting for me (page [19]-137). Before I consider the definition of free universal algebra, I must repeat the preliminary work done by Cohn.

I will start from the page 108.

Cohn starts from category $St$.[2.5] A set is object of the category $St$ and a map is morphism of the category $St$. Let $F$ be functor from category $\mathcal{K}$ to category $St$. Functor $F$ associates with each $a \in Ob\mathcal{K}$ the set $F(a)$ and with each $\mathcal{K}$-morphism
$$\alpha : a \to b$$
the map
$$F(\alpha) : F(a) \to F(b)$$

Cohn has little different definition of universal object (page 109).

Let there exist $u \in Ob\mathcal{K}$, $\rho \in F(u)$ with the property that to each $\xi \in F(a)$ there coresponds unique $\xi' \in \mathrm{Hom}(u, a)$ such that $\xi = \rho\xi'$. Then $u$ is called universal $\mathcal{K}$-object, with universal morhism $\rho$, for the functor $F$.

This definition is little unusual. However, when Cohn introduced category $\mathcal{K}$, he suggested writing $\xi\alpha$ instead of $\xi F(\alpha)$, when $\xi \in F(a)$. It is also unusual that Cohn determines universal object in category $St$. Although, I think this is not important. It is also not clear why $\rho$ is called morphism.

**August 11.** Consider next important for me definition.

DEFINITION 2.10.4. *The category $\mathcal{L}$ is represented*[2.6] *in the category $\mathcal{K}$ if there is covariant functor $F$ from*[2.7] *$\mathcal{L}^\circ \times \mathcal{K}$ to category St.* □

To any $A \in \mathcal{L}$, $a \in \mathcal{K}$, there corresponds a set $F(A, a)$. The map
$$F(\varphi, \alpha) : F(A, a) \to F(B, b)$$
corresponds to morphisms
$$\varphi : B \to A$$
$$\alpha : a \to b$$
We denote the efect of this map on $\xi \in F(A, a)$ by $\varphi\xi\alpha$.

In most applications objects of categories $\mathcal{L}$ and $\mathcal{K}$ will be sets with certain structure, and $F(A, a)$ consists of certain maps from $A$ to $a$. For this reason we generaly refer to elements of $F(A, a)$ as the admissible morphisms of the representation. Instead of the single universal $\mathcal{K}$-object we have one universal

---

[2.5]See example (ii) of category on the page [19]-37.

[2.6]See the definition on the page [19]-109.

[2.7]According to the definition on the page [19]-38, with every category $\mathcal{K}$ we associate the category $\mathcal{K}^\circ$, the oposite of category $\mathcal{K}$, whose morphisms are $\mathcal{K}$-morphisms, but with multiplication

(2.10.1) $$\alpha * \beta = \beta\alpha$$

whenever the right-hand side of the equality (2.10.1) is defined.

$\mathcal{K}$-object for each $\mathcal{L}$-object; this introduces functor from category $\mathcal{L}$ to category $\mathcal{K}$.

**August 13.** I continue to read page [19]-110.

Category $\mathcal{K}$ is subordinate to category $\mathcal{L}$, (we use notation $\mathcal{K} \prec \mathcal{L}$), if there is functor $i$ from category $\mathcal{K}$ to category $\mathcal{L}$ such that
$$\alpha i = \alpha i' \Rightarrow \alpha = \alpha'$$
for any $\mathcal{K}$-morphisms $\alpha, \beta : a \to b$, $a, b \in \text{Ob}\mathcal{K}$. $\mathcal{L}$-object $ai$ corresponding $\mathcal{K}$-object $a$ will be called $\mathcal{L}$-carrier of $\mathcal{K}$-object $a$. Morphism $\alpha i$ will be called $\mathcal{L}$-morphism defined by $\mathcal{K}$-morphism $\alpha$.

In particular, if $\mathcal{K}$ and $\mathcal{L}$ consist of sets with some structure and maps preserving this structure, then $\mathcal{K} \prec \mathcal{L}$ if $\mathcal{L}$-objects have less structure than $\mathcal{K}$-objects, and functor i has the effect of ignoring $\mathcal{K}$-structure, i.e. is a forgetful functor, in MacLane's terminology.

When $\mathcal{K} \prec \mathcal{L}$ and $i$ is the corresponding functor, we can always represent $\mathcal{L}$ in $\mathcal{K}$ by putting
$$F(A, a) = Hom(A, ai) \quad F(\varphi, \alpha) : \xi \to \varphi\xi(\alpha i)$$

**August 16.** On the page [19]-116, Cohn again considers $\Omega$-word algebra. However, Cohn considers here $\Omega$-word algebra differently than on page [19]-79.

STATEMENT 2.10.5. *$\Omega$-algebra is object of the category $(\Omega)$.*[2.8] *A homomorphism of $\Omega$-algebra is morphism of the category $(\Omega)$. The category $(\Omega)$ is subordinate to the category St by the functor which associates with $\Omega$-algebra its carrier. Thus we have representation of category St in $(\Omega)$.* □

Let $X$ be any set. We define $\Omega$-algebra $W(\Omega, X)$, algebra of $\Omega$-rows in $X$, as follows. $\Omega$-row is finite sequence ($n$-tuple for $n \geq 1$) of elements of the disjoint sum $\Omega \sqcup X$. On the set $W(\Omega, X)$ we define $\Omega$-algebra structure by juxtaposition. If $\omega \in \Omega$ and
$$a_i = (a_{i1}, ..., a_{ik_i}) \in W(\Omega, X) \quad i = 1, ..., n \quad a_{ij} \in \Omega \sqcup X$$
then
$$(2.10.2) \qquad a_1...a_n\omega = (a_{11}, ..., a_{1k_1}, a_{21}, ..., a_{nk_n}, \omega)$$

DEFINITION 2.10.6. *The subalgebra $W_\Omega(X)$ of the algebra*[2.9] *$W(\Omega, X)$ generated by the set $X$ is called $\Omega$-word algebra on $X$. Its elements are called $\Omega$-words in $X$, and $X$ is called its alphabet.* □

THEOREM 2.10.7. *$\Omega$-word algebras on sets $X$ and $Y$ are isomorphic*
$$W_\Omega(X) \cong W_\Omega(Y)$$
*iff $X$ is equipotent to $Y$.*

PROOF. See proof of proposition 2.1 on the page [19]-117. □

---
[2.8] See definition of category $(\Omega)$ on the page [19]-50.
[2.9] See definition of algebra $W_\Omega(X)$ on the page [19]-117.

## 2.10. Free Representation

**September 23.** Enough rest. Let us continue. Cohn considers representation of category of $\Omega$-algebras on the page [19]-133. This definition is different from my definition. So I will use full name.

In the statement 2.10.5, we replace category $St$ by category $\mathcal{L}$, $\mathcal{L} \prec St$, and category $(\Omega)$ by category $\mathcal{K}$ of $\Omega$-algebras, $\mathcal{K} \prec \mathcal{L}$. This gives rise to representation of category $\mathcal{L}$ in category $\mathcal{K}$ (called the natural representation). Then Cohn requires

- Any admissible morphism
$$\rho : X \in Ob\mathcal{L} \to A \in Ob\mathcal{K}$$
  determines map (again denoted by $\rho$) from the carrier of $\mathcal{L}$ object $X$ to the carrier of $\mathcal{K}$ object $A$.
- Subalgebra $A_0$ of $\Omega$-algebra $A$ generated by the image of the carrier $X$ under the map $\rho$, is $\mathcal{K}$-algebra.
- Inclusion
$$i : A_0 \to A$$
  is $\mathcal{K}$-homomorphism.
- If $\rho$ is result of cutting down of map $\rho_0$ to $A_0$, so that $\rho = \rho_o \circ i$, then $\rho_0$ is admissible.

Now I recall the definition 2.10.4. The category $\mathcal{L}$ is represented in category $\mathcal{K}$, if there is covariant functor $F$ from $\mathcal{L}^\circ \times \mathcal{K}$ to category $St$. Then the proposition 4.1 follows on the page [19]-133.

THEOREM 2.10.8. *If a category $\mathcal{L}$, $\mathcal{L} \prec St$, is represented in category $\mathcal{K}$, and universal functor $(U, u)$ exists for this representation, then*[2.10] *$U(X)$ is generated by $im\, u$.* □

Now we are ready to return to page [19]-152.

If the universal functor for the representation of $St$ in the category $\mathcal{K}$ of $\Omega$-algebras exists, then, for any set $X$, algebra $U(X)$ is called universal $\mathcal{K}$-algebra over $X$. When universal functor is injective, we identify $X$ with its image in $U(X)$. $U(X)$ is called free $\mathcal{K}$-algebra over $X$ and $X$ is called $\mathcal{K}$-free generating set of algebra $U(X)$.

In other words, the free algebra is algebra with basis.

**September 30.** I think I found a compromise between the definitions 2.10.2 and 2.10.3. I take as an example the theorem [3]-4.1 on the page 135. I have similar theorem [13]-4.7. Actually, this theorem should be formulated in more general form like Lang it did. But now I am looking for inverse theorem.

Before I will continue, I want to be certain about notations. I used notation $\Omega_1(\Omega_2)$ for category of representations of $\Omega_1$-algebra in $\Omega_2$-algebra. However I

---

[2.10]Here for the map
$$\alpha : X \to Y$$
Cohn offers notation $im\, \alpha = \alpha(X)$.

think it will be better to use notation $\Omega_1 \dashrightarrow \Omega_2$. In this case, it will be easier to switch to the corresponding notations for diagrams of representations.

The next remark is about definition of basis. In the definition of the basis of representation $A \dashrightarrow B$ I considered only a set generating universal algebra $B$. This led me to an artificial construction when I defined basis of tower of representations. I think that it is more natural to consider the set of sets generating each algebra. But at this time, I will consider only a set generating the algebra $B$.

Therefore, I will consider the set of $\Omega_2$-algebra $A_2$ containing the set $X$ and some $\Omega_2$-algebra $B_2$. I am interested in representations

$$f : A_1 \dashrightarrow A_2$$

$$g : A_1 \dashrightarrow B_2$$

$$f, g \in A_1 \dashrightarrow \Omega_2$$

and reduced morphism of representations

$$h : A_2 \to B_2$$

for which the image $h(X)$ is given. I am interested in such representation $f$ for which reduced morphism $h$ is uniquely defined. Can I express this construction on category language?

**November 27.** I tried to consider three definitions of free representation. I usually use the definition 2.10.1. The following definition is attempt to give definition following definition by Leng.

DEFINITION 2.10.9. *Let $C \subseteq A$. The map*

$$f : A \to B$$

*is called extension of the map*

$$g : C \to B$$

*if $x \in C \Rightarrow f(x) = g(x)$.* □

DEFINITION 2.10.10. *The representation*

$$f : A_1 \dashrightarrow A_2$$

*of $\Omega_1$-algebra $A_1$ in $\Omega_2$-algebra $A_2$ is called* **free representation**, *if there exist the set $X \subseteq A_2$ such that the map*

$$p : X \to B_2$$

*of the set $X$ into $\Omega_2$-algebra $B_2$ has unique extension*

$$r : A_1 \to B_2$$

*which is reduced morphism of representation $f$ into representation*

$$g : A_1 \dashrightarrow B_2$$

□

## 2.10. Free Representation

In fact, I should have started from the definition 2.10.2. But I had met the problem to define free object.

And finally, the following definition was inspired by the definition of free module.

DEFINITION 2.10.11. *The representation*
$$f: A_1 \relbar\joinrel\twoheadrightarrow A_2$$
*is called free representation if this representation has a basis.* □

**November 29.** So my task is to prove that these definitions are equivalent. Based on the definition 2.10.10, I started from the following theorem.

THEOREM 2.10.12. *Let the representation*
$$f: A_1 \relbar\joinrel\twoheadrightarrow A_2$$
*is free according to the definition 2.10.10. Then the set $X$ is generating set of representation $f$.*

PROOF. Consider the map
$$p: x \in X \to x \in A_2$$
According to theorems [10]-2.2.7, [10]-2.6.4, the map
$$r: A_2 \to A_2$$
has form $r(x) = x$ on the set $J_f(X)$. However, if the set $A_2 \setminus J_f(X)$ is not empty, then the map $r$ has arbitrary form on the set $A_2 \setminus J_f(X)$. Therefore, the set $X$ is generating set of representation $f$. □

After this I tried to prove the following theorem.

THEOREM 2.10.13. *Let*
$$f: A_1 \relbar\joinrel\twoheadrightarrow A_2$$
*be free representation according to the definition 2.10.1. Then the representation $f$ is free according to the definition 2.10.10.*

PROOF. The following idea seemed obvious to me to begin the proof.

Let $X$ be generating set of representation $f$. Consider maps
$$p_1: a_2 \in X \to f(a_1)(a_2) \in A_2$$
$$p_2: a_2 \in X \to f(b_1)(a_2) \in A_2$$
Let
$$p_1(x) = p_2(x)$$
The reason of this choice is simple. It is easy to extend such map from the set $C$ to set $A_2$. Because the extension is unique, then I can say that $a_1 = b_1$ (I still have to be careful here). In order not to repeat the same reasoning twice, I decided to prove the following lemma.

LEMMA 2.10.14. *Let $X$ be generating set of representation $f$. Let the map*
$$p : a_2 \in X \to f(a_1)(a_2) \in A_2$$
*has extension*
$$r : A_1 \to B_2$$
*which is reduced endomorphism of representation $f$. Then the map $r$ has the following form*
$$r(a_2) = f(a_1)(a_2)$$

PROOF. Based on the theorem [10]-2.6.4, we will prove lemma by induction on a chain of sets $X_k$ generating representation $f$.

The lemma is true for the set $X_0 = X$, because this is the main prerequisite of the lemma.

Let the lemma be true for the set $X_k$.

- Let $a_2 \in X_k$. According to the statement [10]-2.6.4.4, $f(b_1)(a_2) \in X_{k+1}$.
- 

⊙

I stopped here. I cannot write
$$f(a_1)(f(b_1)(a_2)) = f(b_1)(f(a_1)(a_2))$$
Therefore, I cannot extend the map $p$ to reduced endomorphism. □

In this place I decided to revise the theorem 2.10.13. I have remembered about the theorem [10]-2.6.11. Against the background of this theorem, the theorem 2.10.13 seems meaningless. If $X$ is generating set of the representation, then an extension of the map $p$ is unique. For the present, I decided to be satisfied with the definitions 2.10.10 and 2.10.11.

## 2.11. System of Linear Equations

**August 18: Quaternion Algebra.** I decided to take a short break with universal algebras. I want to consider system of linear equations over quaternion algebra. I will begin to study the system of linear equations over quaternion algebra with a relatively simple example.

EXAMPLE 2.11.1. *Consider the system of linear equations over quaternion algebra*

(2.11.1)
$$\begin{aligned} ix^1 j + jx^2 k &= i + j \\ kx^1 i + ix^2 j &= i - j \end{aligned}$$

*We represent the system of linear equations* (2.11.1) *in the following way*

(2.11.2)
$$\begin{aligned} (i \otimes j) \circ x^1 + (j \otimes k) \circ x^2 &= i + j \\ (k \otimes i) \circ x^1 + (i \otimes j) \circ x^2 &= i - j \end{aligned}$$

(2.11.3) $$\begin{pmatrix} i\otimes j & j\otimes k \\ k\otimes i & i\otimes j \end{pmatrix} {}_\circ^\circ \begin{pmatrix} cx^1 \\ x^2 \end{pmatrix} = \begin{pmatrix} i+j \\ i-j \end{pmatrix}$$

Now we calculate ${}_\circ^\circ$-quasideterminant of matrix
$$a = \begin{pmatrix} i\otimes j & j\otimes k \\ k\otimes i & i\otimes j \end{pmatrix}$$

According to the equality [11]-(2.3.10), we get

(2.11.4)
$$\begin{aligned}
\det(a,{}_\circ^\circ)^1_1 &= a^1_1 - a^1_{[1]} {}_\circ^\circ (a^{[1]}_{[1]})^{-1} {}_\circ^\circ a^{[1]}_1 = a^1_1 - a^1_2 \circ (a^2_2)^{-1} \circ a^2_1 \\
&= i\otimes j - (j\otimes k)\circ(i\otimes j)^{-1}\circ(k\otimes i) \\
&= i\otimes j - (j\otimes k)\circ(i\otimes j)\circ(k\otimes i) = i\otimes j - (-1)\otimes(-1) \\
&= i\otimes j - 1\otimes 1
\end{aligned}$$

(2.11.5)
$$\begin{aligned}
\det(a,{}_\circ^\circ)^1_2 &= a^1_2 - a^1_{[2]} {}_\circ^\circ (a^{[1]}_{[2]})^{-1} {}_\circ^\circ a^{[1]}_2 = a^1_2 - a^1_1 \circ (a^2_1)^{-1} \circ a^2_2 \\
&= j\otimes k - (i\otimes j)\circ(k\otimes i)^{-1}\circ(i\otimes j) \\
&= j\otimes k - (i\otimes j)\circ(k\otimes i)\circ(i\otimes j) = j\otimes k - (-k)\otimes i \\
&= j\otimes k + k\otimes i
\end{aligned}$$

(2.11.6)
$$\begin{aligned}
\det(a,{}_\circ^\circ)^2_1 &= a^2_1 - a^2_{[1]} {}_\circ^\circ (a^{[2]}_{[1]})^{-1} {}_\circ^\circ a^{[2]}_1 = a^2_1 - a^2_2 \circ (a^1_2)^{-1} \circ a^1_1 \\
&= k\otimes i - (i\otimes j)\circ(j\otimes k)^{-1}\circ(i\otimes j) \\
&= k\otimes i - (i\otimes j)\circ(j\otimes k)\circ(i\otimes j) = k\otimes i - (-j)\otimes k \\
&= k\otimes i + j\otimes k
\end{aligned}$$

(2.11.7)
$$\begin{aligned}
\det(a,{}_\circ^\circ)^2_2 &= a^2_2 - a^2_{[2]} {}_\circ^\circ (a^{[2]}_{[2]})^{-1} {}_\circ^\circ a^{[2]}_2 = a^2_2 - a^2_1 \circ (a^1_1)^{-1} \circ a^1_2 \\
&= i\otimes j - (k\otimes i)\circ(i\otimes j)^{-1}\circ(j\otimes k) \\
&= i\otimes j - (k\otimes i)\circ(i\otimes j)\circ(j\otimes k) = i\otimes j - (-1)\otimes(-1) \\
&= i\otimes j - 1\otimes 1
\end{aligned}$$

□

**August 19.** In order to finish the calculations in the example 2.11.1, we need to answer the next question. Is there a tensor $a^{-1}$ for given tensor $a \in A \otimes A$ and how we can find the tensor $a^{-1}$, if it exists?

There are 2 ways to answer this question. We may consider standard representation of tensors $a$ and $a^{-1}$. Or, keeping in mind that tensor $a$ corresponds to linear map of quaternion algebra, we may consider inverse map.

We will start from standard representation of tensor $a$.

THEOREM 2.11.2. *Let $C^k_{ij}$ be structural constants of D-algebra A. Let*

(2.11.8) $$f = f^{ij} e_i \otimes e_j$$

be standard representation of the tensor $f \in A \otimes A$ and
(2.11.9) $$g = g^{ij} e_i \otimes e_j$$
be standard representation of the tensor $g \in A \otimes A$. Then the standard representation of product of tensors
(2.11.10) $$f \circ g = (f \circ g)^{pq} e_p \otimes e_q$$
satisfies to the equality
(2.11.11) $$(f \circ g)^{pq} = f^{ij} g^{kl} C^p_{ik} C^q_{lj}$$

PROOF. The equality
(2.11.12) $$f \circ g = f^{ij} g^{kl} (e_i e_k) \otimes (e_l e_j)$$
follows from equalities (2.11.8), (2.11.9). The equality
(2.11.13) $$f \circ g = f^{ij} g^{kl} C^p_{ik} C^q_{lj} e_p \otimes e_q$$
follows from equalities [11]-(3.2.5), (2.11.12). The equality (2.11.11) follows from equalities (2.11.10), (2.11.13). □

THEOREM 2.11.3. Let $C^k_{ij}$ be structural constants of D-algebra A. Let
$$f = f^{ij} e_i \otimes e_j$$
be standard representation of the tensor $f \in A \otimes A$. Let there exist tensor $g = f^{-1}$ and
$$g = g^{ij} e_i \otimes e_j$$
be standard representation of the tensor g. Then standard components of the tensor g satisfy to the system of linear equations
(2.11.14) $$\begin{aligned} f^{ij} C^0_{ik} C^0_{lj} g^{kl} &= 1 \\ f^{ij} C^p_{ik} C^q_{lj} g^{kl} &= 0 \quad p+q > 0 \end{aligned}$$

PROOF. The theorem follows from the equality
$$g \circ h = e_0 \otimes e_0$$
and from the theorem 2.11.2. □

**August 20.** Calculation of the coefficients of the system of linear equations (2.11.14) is wearisome exercise. Even I will will use computer for this task, A system of linear equations of order 16 looks rather cumbersome. This is why I deciced to start from solving of specific problem.

EXAMPLE 2.11.4. Based on the theorem 2.11.3, we can continue study of example 2.11.1. We have tensors
$$f_1 = -1 \otimes 1 + i \otimes j$$
$$f_2 = k \otimes i + j \otimes k$$

## 2.11. System of Linear Equations

We want to find tensors $g_1 = f_1^{-1}$ and $g_2 = f_2^{-1}$. The system of linear equations (2.11.14) for tensor $g_1$ has form

$$(2.11.15) \begin{cases} -C^0_{0k}C^0_{l0}g^{kl} + C^0_{1k}C^0_{l2}g^{kl} = 1 & -C^2_{0k}C^0_{l0}g^{kl} + C^2_{1k}C^0_{l2}g^{kl} = 0 \\ -C^0_{0k}C^1_{l0}g^{kl} + C^0_{1k}C^1_{l2}g^{kl} = 0 & -C^2_{0k}C^1_{l0}g^{kl} + C^2_{1k}C^1_{l2}g^{kl} = 0 \\ -C^0_{0k}C^2_{l0}g^{kl} + C^0_{1k}C^2_{l2}g^{kl} = 0 & -C^2_{0k}C^2_{l0}g^{kl} + C^2_{1k}C^2_{l2}g^{kl} = 0 \\ -C^0_{0k}C^3_{l0}g^{kl} + C^0_{1k}C^3_{l2}g^{kl} = 0 & -C^2_{0k}C^3_{l0}g^{kl} + C^2_{1k}C^3_{l2}g^{kl} = 0 \\ -C^1_{0k}C^0_{l0}g^{kl} + C^1_{1k}C^0_{l2}g^{kl} = 0 & -C^3_{0k}C^0_{l0}g^{kl} + C^3_{1k}C^0_{l2}g^{kl} = 0 \\ -C^1_{0k}C^1_{l0}g^{kl} + C^1_{1k}C^1_{l2}g^{kl} = 0 & -C^3_{0k}C^1_{l0}g^{kl} + C^3_{1k}C^1_{l2}g^{kl} = 0 \\ -C^1_{0k}C^2_{l0}g^{kl} + C^1_{1k}C^2_{l2}g^{kl} = 0 & -C^3_{0k}C^2_{l0}g^{kl} + C^3_{1k}C^2_{l2}g^{kl} = 0 \\ -C^1_{0k}C^3_{l0}g^{kl} + C^1_{1k}C^3_{l2}g^{kl} = 0 & -C^3_{0k}C^3_{l0}g^{kl} + C^3_{1k}C^3_{l2}g^{kl} = 0 \end{cases}$$

The system of linear equations

$$(2.11.16) \begin{cases} -g^{00} + g^{12} = 1 & -C^2_{0k}C^0_{l0}g^{kl} + C^2_{1k}C^0_{l2}g^{kl} = 0 \\ -g^{01} + g^{13} = 0 & -C^2_{0k}C^1_{l0}g^{kl} + C^2_{1k}C^1_{l2}g^{kl} = 0 \\ -g^{02} - g^{10} = 0 & -C^2_{0k}C^2_{l0}g^{kl} + C^2_{1k}C^2_{l2}g^{kl} = 0 \\ -g^{03} + g^{11} = 0 & -C^2_{0k}C^3_{l0}g^{kl} + C^2_{1k}C^3_{l2}g^{kl} = 0 \\ -g^{10} - g^{02} = 0 & -C^3_{0k}C^0_{l0}g^{kl} + C^3_{1k}C^0_{l2}g^{kl} = 0 \\ -g^{11} - g^{03} = 0 & -C^3_{0k}C^1_{l0}g^{kl} + C^3_{1k}C^1_{l2}g^{kl} = 0 \\ -g^{12} + g^{00} = 0 & -C^3_{0k}C^2_{l0}g^{kl} + C^3_{1k}C^2_{l2}g^{kl} = 0 \\ -g^{13} + g^{01} = 0 & -C^3_{0k}C^3_{l0}g^{kl} + C^3_{1k}C^3_{l2}g^{kl} = 0 \end{cases}$$

follows from the system of linear equations (2.11.15) and from the equality

$$(2.11.17) \begin{array}{llll} C^0_{00} = 1 & C^1_{01} = 1 & C^2_{02} = 1 & C^3_{03} = 1 \\ C^1_{10} = 1 & C^0_{11} = -1 & C^3_{12} = 1 & C^2_{13} = -1 \\ C^2_{20} = 1 & C^3_{21} = -1 & C^0_{22} = -1 & C^1_{23} = 1 \\ C^3_{30} = 1 & C^2_{31} = 1 & C^1_{32} = -1 & C^0_{33} = -1 \end{array}$$

In the system of equations (2.11.16), from a comparison of the equations highlighted in blue color, it follows that the system of equations (2.11.16) has no solution.

*The system of linear equations* (2.11.14) *for tensor $g_2$ has form*

(2.11.18)
$$\begin{cases} C^0_{3k}C^0_{l1}g^{kl} + C^0_{2k}C^0_{l3}g^{kl} = 1 & C^2_{3k}C^0_{l1}g^{kl} + C^2_{2k}C^0_{l3}g^{kl} = 0 \\ C^0_{3k}C^1_{l1}g^{kl} + C^0_{2k}C^1_{l3}g^{kl} = 0 & C^2_{3k}C^1_{l1}g^{kl} + C^2_{2k}C^1_{l3}g^{kl} = 0 \\ C^0_{3k}C^2_{l1}g^{kl} + C^0_{2k}C^2_{l3}g^{kl} = 0 & C^2_{3k}C^2_{l1}g^{kl} + C^2_{2k}C^2_{l3}g^{kl} = 0 \\ C^0_{3k}C^3_{l1}g^{kl} + C^0_{2k}C^3_{l3}g^{kl} = 0 & C^2_{3k}C^3_{l1}g^{kl} + C^2_{2k}C^3_{l3}g^{kl} = 0 \\ C^1_{3k}C^0_{l1}g^{kl} + C^1_{2k}C^0_{l3}g^{kl} = 0 & C^3_{3k}C^0_{l1}g^{kl} + C^3_{2k}C^0_{l3}g^{kl} = 0 \\ C^1_{3k}C^1_{l1}g^{kl} + C^1_{2k}C^1_{l3}g^{kl} = 0 & C^3_{3k}C^1_{l1}g^{kl} + C^3_{2k}C^1_{l3}g^{kl} = 0 \\ C^1_{3k}C^2_{l1}g^{kl} + C^1_{2k}C^2_{l3}g^{kl} = 0 & C^3_{3k}C^2_{l1}g^{kl} + C^3_{2k}C^2_{l3}g^{kl} = 0 \\ C^1_{3k}C^3_{l1}g^{kl} + C^1_{2k}C^3_{l3}g^{kl} = 0 & C^3_{3k}C^3_{l1}g^{kl} + C^3_{2k}C^3_{l3}g^{kl} = 0 \end{cases}$$

*The system of linear equations*

(2.11.19)
$$\begin{cases} g^{31} + g^{23} = 1 & C^2_{3k}C^0_{l1}g^{kl} + C^2_{2k}C^0_{l3}g^{kl} = 0 \\ -g^{30} - g^{22} = 0 & C^2_{3k}C^1_{l1}g^{kl} + C^2_{2k}C^1_{l3}g^{kl} = 0 \\ -g^{33} + g^{21} = 0 & C^2_{3k}C^2_{l1}g^{kl} + C^2_{2k}C^2_{l3}g^{kl} = 0 \\ g^{32} - g^{20} = 0 & C^2_{3k}C^3_{l1}g^{kl} + C^2_{2k}C^3_{l3}g^{kl} = 0 \\ g^{21} - g^{33} = 0 & C^3_{3k}C^0_{l1}g^{kl} + C^3_{2k}C^0_{l3}g^{kl} = 0 \\ -g^{20} + g^{32} = 0 & C^3_{3k}C^1_{l1}g^{kl} + C^3_{2k}C^1_{l3}g^{kl} = 0 \\ -g^{23} - g^{31} = 0 & C^3_{3k}C^2_{l1}g^{kl} + C^3_{2k}C^2_{l3}g^{kl} = 0 \\ g^{22} + g^{30} = 0 & C^3_{3k}C^3_{l1}g^{kl} + C^3_{2k}C^3_{l3}g^{kl} = 0 \end{cases}$$

*follows from the system of linear equations* (2.11.18) *and from the equality* (2.11.17). *In the system of equations* (2.11.19), *from a comparison of the equations highlighted in blue color, it follows that the system of equations* (2.11.19) *has no solution.* □

**August 21.** There is probability that randomly chosen tensor $a \in A \otimes A$ does not have inverse. However when two different tensors show absolutely identical behavior, the question about error arises.

After I finished the calculations in the example 2.11.4, I am ready to return to the system of linear equations (2.11.14). I will start from the first equation. $C^0_{ik}$ is different from 0 when $i = k$. Therefore, we can write the first equation by the following way

(2.11.20) $$f^{00}g^{00} - f^{0\alpha}g^{0\alpha} - f^{\beta 0}g^{\beta 0} + f^{\beta\alpha}g^{\beta\alpha} = 1$$
$$\alpha, \beta = 1, 2, 3$$

Now let us see if we can reproduce the equation (2.11.20) in second row of the system of equations (2.11.14). Indexes $p$, $q$ are number of second equation. We

## 2.11. System of Linear Equations

can write the second equation by the following way

(2.11.21) $\quad f^{ij}C^p_{i0}C^q_{0j}g^{00} + f^{ij}C^p_{i0}C^q_{\alpha j}g^{0\alpha} + f^{ij}C^p_{i\beta}C^q_{0j}g^{\beta 0} + f^{ij}C^p_{i\beta}C^q_{\alpha j}g^{\beta\alpha} = 0$

Now I see that irreversibility of tensors $f_1$, $f_2$ is inescapable.

I see a lot of interesting questions. In particular, how does this statement affect a solving of system of linear equations. At the same time, I am ready now to consider any system of linear equations.

**August 22.** Before moving on, I can verify the equality (2.11.21) when $p = 1$, $q = 2$

(2.11.22) $\quad f^{ij}C^1_{i0}C^2_{0j}g^{00} + f^{ij}C^1_{i0}C^2_{\alpha j}g^{0\alpha} + f^{ij}C^1_{i\beta}C^2_{0j}g^{\beta 0} + f^{ij}C^1_{i\beta}C^2_{\alpha j}g^{\beta\alpha} = 0$

(2.11.23) $\quad f^{12}C^1_{00}C^2_{02}g^{00} + f^{1j}C^1_{10}C^2_{\alpha j}g^{0\alpha} + f^{ij}C^1_{i\beta}C^2_{0j}g^{\beta 0} + f^{ij}C^1_{i\beta}C^2_{\alpha j}g^{\beta\alpha} = 0$

Indeed, it was necessary to guess so. I will consider it as a gift of fate; otherwise I could stay in happy ignorance.

It is interesting what are matrices of maps $f_1$, $f_2$. According to the theorem [11]-4.3.3, matrices of maps $f_1$, $f_2$ are

(2.11.24)
$$\begin{pmatrix} f^0_0 & f^0_1 & f^0_2 & f^0_3 \\ f^1_1 & -f^1_0 & f^1_3 & -f^1_2 \\ f^2_2 & -f^2_3 & -f^2_0 & f^2_1 \\ f^3_3 & f^3_2 & -f^3_1 & -f^3_0 \end{pmatrix}$$
$$= \begin{pmatrix} 1 & -1 & -1 & -1 \\ 1 & -1 & 1 & 1 \\ 1 & 1 & -1 & 1 \\ 1 & 1 & 1 & -1 \end{pmatrix} \begin{pmatrix} -1 & 0 & 0 & 0 \\ 0 & 0 & 0 & -1 \\ 0 & 0 & 0 & 0 \\ 0 & 0 & 0 & 0 \end{pmatrix}$$

(2.11.25)
$$\begin{pmatrix} f^0_0 & f^0_1 & f^0_2 & f^0_3 \\ f^1_1 & -f^1_0 & f^1_3 & -f^1_2 \\ f^2_2 & -f^2_3 & -f^2_0 & f^2_1 \\ f^3_3 & f^3_2 & -f^3_1 & -f^3_0 \end{pmatrix}$$
$$= \begin{pmatrix} 1 & -1 & -1 & -1 \\ 1 & -1 & 1 & 1 \\ 1 & 1 & -1 & 1 \\ 1 & 1 & 1 & -1 \end{pmatrix} \begin{pmatrix} 0 & 0 & 0 & 0 \\ 0 & 0 & 0 & 0 \\ 0 & -1 & 0 & 0 \\ 0 & 0 & -1 & 0 \end{pmatrix}$$

Matrices of maps $f_1$, $f_2$ are singular. But the same thing I can say, if $f = 1 \otimes 1$. It is necessary to understand the equality [11]-(3.8.4).

**August 23.** So I return to the following theorem.

THEOREM 2.11.5. *Let $\overline{\overline{e}}$ be basis of the free finite dimensional D-module A. Let $C_{kl}^{p}$ be structural constants of D-algebra A. Let $\overline{\overline{F}}$ be the basis of left $A \otimes A$-module $\mathcal{L}(D; A \to A)$ and $F_{k \cdot i}^{\ \ j}$ be coordinates of map $F_k$ with respect to basis $\overline{\overline{e}}$. Coordinates $f_l^k$ of the map $f \in \mathcal{L}(D; A \to A)$ and its standard components $f^{k \cdot ij}$ are connected by the equation*

$$(2.11.26) \qquad f_l^k = f^{k \cdot ij} F_{kl}^{\ \ m} C_{im}^p C_{pj}^k$$

PROOF. Relative to basis $\overline{\overline{e}}$, linear maps $f$ and $F_k$ have form

$$(2.11.27) \qquad f \circ x = f_j^i x^j e_i$$

$$(2.11.28) \qquad F_k \circ x = F_{k \cdot j}^{\ \ i} x^j e_i$$

The equality

$$(2.11.29) \qquad \begin{aligned} f_l^k x^l e_k &= f^{k \cdot ij} e_i F_{k \cdot l}^{\ \ m} x^l e_m e_j \\ &= f^{k \cdot ij} F_{k \cdot l}^{\ \ m} x^l C_{im}^p C_{pj}^k e_k \end{aligned}$$

follows from equalities

$$(2.11.30) \qquad f = f^{k \cdot ij} (e^i \otimes e^j) \circ F_k = f^{k \cdot ij} e^i F_k e^j$$

and from equalities (2.11.27), (2.11.28). Since vectors $e_k$ are linear independent and $x^i$ are arbitrary, then the equation (2.11.26) follows from the equation (2.11.29). □

The theorem is very simple. However it has somewhere underwater stone.

**August 25.** Without loss of generality, let dimension of left $A \otimes A$-module $\mathcal{L}(D; A \to A)$ be 1. Then the equality (2.11.29) gets the following form

$$(2.11.31) \qquad f_l^k x^l e_k = f^{ij} e_i x^m e_m e_j = f^{ij} x^m C_{im}^p C_{pj}^k e_k$$

I will consider the most simple linear map

$$(2.11.32) \qquad f \circ x = ax \quad a \in A$$

The standard representation of the map $f$ has form

$$(2.11.33) \qquad f \circ x = a^i (e_i \otimes 1) \circ (x^j e_j) = a^i x^j e_i e_j = a^i x^j C_{ij}^k e_k$$

From the equality

$$f \circ x = C_{ij}^k a^i x^j e_k$$

it follows that

$$(2.11.34) \qquad f_j^k = \frac{\partial f^k}{\partial x^j} = C_{ij}^k a^i$$

## 2.11. System of Linear Equations

At the same time, the equality

(2.11.35)
$$f_j^k = a^i C_{ij}^k$$

follows from the equality (2.11.33). Equalities (2.11.34), (2.11.35) are the same. This is interesting. It seems that there is nothing wrong in the theorem 2.11.5.

Back to the equality (2.11.31)

(2.11.36)
$$f_l^k = f^{ij} C_{il}^p C_{pj}^k$$

Since there are different from 0 only standard components $f^{i0} = a^i$, then the equality

(2.11.37)
$$f_l^k = a^i C_{il}^p C_{p0}^k = a^i C_{il}^k$$

follows from the equality (2.11.36). If only $a^0$ is different from 0, then the equality

(2.11.38)
$$f_l^k = a^0 C_{0l}^p = a^i \delta_l^k$$

follows from the equality (2.11.37). And I expected this.

Actually, there is no problem. This is simply a consequence of inattention. Entries of matrix $f$ in left part of equalities (2.11.24), (2.11.25) are arranged in an order different from the standard.

I will finish the equality (2.11.24) for the tensor

$$f_1 = -1 \otimes 1 + i \otimes j$$

(2.11.39)
$$\begin{pmatrix} f_0^0 & f_1^0 & f_2^0 & f_3^0 \\ f_1^1 & -f_0^1 & f_3^1 & -f_2^1 \\ f_2^2 & -f_3^2 & -f_0^2 & f_1^2 \\ f_3^3 & f_2^3 & -f_1^3 & -f_0^3 \end{pmatrix}$$

$$= \begin{pmatrix} 1 & -1 & -1 & -1 \\ 1 & -1 & 1 & 1 \\ 1 & 1 & -1 & 1 \\ 1 & 1 & 1 & -1 \end{pmatrix} \begin{pmatrix} -1 & 0 & 0 & 0 \\ 0 & 0 & 0 & -1 \\ 0 & 0 & 0 & 0 \\ 0 & 0 & 0 & 0 \end{pmatrix}$$

$$= \begin{pmatrix} -1 & 0 & 0 & 1 \\ -1 & 0 & 0 & 1 \\ -1 & 0 & 0 & -1 \\ -1 & 0 & 0 & -1 \end{pmatrix}$$

The equality

(2.11.40) $$f_1 = \begin{pmatrix} -1 & 0 & 0 & 1 \\ 0 & -1 & -1 & 0 \\ 0 & -1 & -1 & 0 \\ 1 & 0 & 0 & -1 \end{pmatrix}$$

follows from the equality (2.11.39). From the equality (2.11.40), it follows that rank of the matrix $f_1$ is 2.

I will finish the equality (2.11.25) for the tensor

$$f_2 = k \otimes i + j \otimes k$$

(2.11.41)
$$\begin{pmatrix} f_0^0 & f_1^0 & f_2^0 & f_3^0 \\ f_1^1 & -f_0^1 & f_3^1 & -f_2^1 \\ f_2^2 & -f_3^2 & -f_0^2 & f_1^2 \\ f_3^3 & f_2^3 & -f_1^3 & -f_0^3 \end{pmatrix}$$
$$= \begin{pmatrix} 1 & -1 & -1 & -1 \\ 1 & -1 & 1 & 1 \\ 1 & 1 & -1 & 1 \\ 1 & 1 & 1 & -1 \end{pmatrix} \begin{pmatrix} 0 & 0 & 0 & 0 \\ 0 & 0 & 0 & 0 \\ 0 & -1 & 0 & 0 \\ 0 & 0 & -1 & 0 \end{pmatrix}$$
$$= \begin{pmatrix} 0 & 1 & 1 & 0 \\ 0 & -1 & -1 & 0 \\ 0 & 1 & -1 & 0 \\ 0 & -1 & 1 & 0 \end{pmatrix}$$

The equality

(2.11.42) $$f_2 = \begin{pmatrix} 0 & 1 & 1 & 0 \\ 1 & 0 & 0 & -1 \\ 1 & 0 & 0 & -1 \\ 0 & -1 & -1 & 0 \end{pmatrix}$$

follows from the equality (2.11.41). From the equality (2.11.42), it follows that rank of the matrix $f_2$ is 2.

## 2.12. Division of polynomials

**October 6.** I returned to the book [18]. It bothers me how well I performed the division operation in the preface.

I will statrt from polynomial
$$p(x) = 2(x-i)(x-j) + (x-j)(x-i) = 3x^2 - 2ix - 2xj - jx - xi + k$$
and divide this polynomial over the polynomial
(2.12.1) $$r(x) = x - i$$
The most simple form of division has form
(2.12.2) $$p(x) = (2 \otimes (x-j)) \circ r(x) + ((x-j) \otimes 1) \circ r(x)$$
I will write the equality (2.12.2) as
$$p(x) = s_1(x) \circ r(x)$$
where
$$s_1(x) = 2 \otimes x - 2 \otimes j + x \otimes 1 - j \otimes 1$$
It is evident that
$$s_1(j) = 0 \otimes 0$$
However, in the book [18], I used division algorithm and got different result. In the paper, I used equality x=r(x)+i So division algorithm got form
$$p(x) = 3(r(x) + i)x - 2ix - 2xj - jx - xi + k$$
$$= 3r(x)x + 3ix - 2ix - 2xj - jx - xi + k$$
$$= 3r(x)x + ix - 2xj - jx - xi + k$$
$$= 3r(x)x + i(r(x) + i) - 2(r(x) + i)j - j(r(x) + i) - (r(x) + i)i + k$$
$$= 3r(x)x + ir(x) - 1 - 2r(x)j - 2ij - jr(x) - ji - r(x)i - ii + k$$
$$= 3r(x)x + ir(x) - 1 - 2r(x)j - 2k - jr(x) + k - r(x)i + 1 + k$$
$$= 3r(x)x + ir(x) - 2r(x)j - jr(x) - r(x)i$$
$$= (3 \otimes x + i \otimes 1 - 2 \otimes j - j \otimes 1 - 1 \otimes i) \circ r(x)$$
where quotient of polynomial $p(x)$ divided by polynomial $r(x)$ is the tensor
$$s(x) = 3 \otimes x + i \otimes 1 - 2 \otimes j - j \otimes 1 - 1 \otimes i$$
It is evident that
$$s(j) = 1 \otimes (j - i) - (j - i) \otimes 1 \neq 0 \otimes 0$$
Somewhere there is error here.

There is no error. But it turns out the result of the division depends on whether I will write
$$3x^2 = (3 \otimes x) \circ x$$
or
$$3x^2 = (2 \otimes x + x \otimes 1) \circ x$$

**October 7.** Initially this statement looks shocking. However, on reflection, I recollected that it is possible infinity many roots for quadratic equation. And this should be reflected in the decomposition of polynomial of second degree over linear factors.

So I decided to look how spectrum of quotients arise when I divide polynomials. I decided to write down polynomial $p(x)$ in the following form
$$p(x) = ((3-a)(x \otimes 1) + a \otimes x) \circ x - 2ix - 2xj - jx - xi + k$$
Then we can present polynomial $p(x)$ in the following form
$$p(x) = (3-a)(r(x) + i)x + ax(r(x) + i) - 2ix - 2xj - jx - xi + k$$
$$= (3-a)r(x)x + (3-a)ix + axr(x) + axi - 2ix - 2xj - jx - xi + k$$
$$= (3-a)r(x)x + axr(x) + (3-a)i(r(x) + i)$$
$$+ a(r(x) + i)i - 2i(r(x) + i) - 2(r(x) + i)j - j(r(x) + i) - (r(x) + i)i + k$$
$$= (3-a)r(x)x + axr(x) + 3ir(x) - 3 - air(x) + a$$
$$+ ar(x)i - a - 2ir(x) + 2 - 2r(x)j - 2k - jr(x) + k - r(x)i + 1 + k$$
$$= (3-a)r(x)x + axr(x) + 3ir(x) - air(x)$$
$$+ ar(x)i - 2ir(x) - 2r(x)j - jr(x) - r(x)i$$
Therefore
$$p(x) = s_2(a, x) \circ r(x)$$
where
$$s_2(a, x) = (3-a) \otimes x + ax \otimes 1 + (3-a)i \otimes 1$$
$$+ a \otimes i - 2i \otimes 1 - 2 \otimes j - j \otimes 1 - 1 \otimes i$$
$$= (3-a) \otimes x + (ax - j - 2i + (3-a)i) \otimes 1 + (a-1) \otimes i - 2 \otimes j$$
$$= (3-a) \otimes x + (ax - j + i - ai) \otimes 1 + (a-1) \otimes i - 2 \otimes j$$
It is easy to see that
$$s_2(1, x) = 2 \otimes x + (x - j - 2i + 2i) \otimes 1 + (1-1) \otimes i - 2 \otimes j$$
$$= 2 \otimes (x - j) + (x - j) \otimes 1$$
and therefore
$$s_2(1, j) = 2 \otimes (j - j) + (j - j) \otimes 1 = 2 \otimes 0 + 0 \otimes 1$$
Now I face two problems.

QUESTION 2.12.1. *Let the polynomial $r(x)$ of degree 1 is divisor of the polynomial $p(x)$ of degree 2. What is the set of tensors $s(x) \in A \otimes A$ such that*
(2.12.3) $$p(x) = s(x) \circ r(x)$$
☐

QUESTION 2.12.2. *For given tensor $s(x) \in A \otimes A$, what is the set of $A$-numbers such that $s(x) = 0$?*
☐

**October 8.** I think the answer on the question 2.12.2 is more simple. Since I can identify the tensor $s(x) \in A \otimes A$ and corresponding matrix $S(x)$ of linear map, then I have to write down the matrix $S(x)$ and find $x$ for which $S(x) = 0$. Since $S(x)$ depends on $x$ linearly, then I need to solve system of linear equations with respect to coordinates of $A$-number $x$.

EXAMPLE 2.12.3. *Consider the tensor*

$$(2.12.4) \quad s_2(a,x) = (3-a) \otimes x + (ax - j + i - ai) \otimes 1 + (a-1) \otimes i - 2 \otimes j$$

*where $a \in R$, $s_2(a,x) \in H \otimes H$. According to theorems [15]-3.1, [15]-3.2, the matrix $S(a,x)$ has following form*

$$(2.12.5) \quad S_2(a,x) = (3-a)E_r(x) + E_l(ax - j + i - ai) + (a-1)E_r(i) - 2E_r(j)$$

$$(2.12.6) \quad E_r(x) = \begin{pmatrix} x^0 & -x^1 & -x^2 & -x^3 \\ x^1 & x^0 & x^3 & -x^2 \\ x^2 & -x^3 & x^0 & x^1 \\ x^3 & x^2 & -x^1 & x^0 \end{pmatrix}$$

$$(2.12.7) \quad \begin{aligned} &E_l(ax - j + i - ai) \\ &= \begin{pmatrix} ax^0 & -ax^1 - 1 + a & -ax^2 + 1 & -ax^3 \\ ax^1 + 1 - a & ax^0 & -ax^3 & ax^2 - 1 \\ ax^2 - 1 & ax^3 & ax^0 & -ax^1 - 1 + a \\ ax^3 & -ax^2 + 1 & ax^1 + 1 - a & ax^0 \end{pmatrix} \end{aligned}$$

$$(2.12.8) \quad E_r(i) = \begin{pmatrix} 0 & -1 & 0 & 0 \\ 1 & 0 & 0 & 0 \\ 0 & 0 & 0 & 1 \\ 0 & 0 & -1 & 0 \end{pmatrix}$$

$$(2.12.9) \quad E_r(j) = \begin{pmatrix} 0 & 0 & -1 & 0 \\ 0 & 0 & 0 & -1 \\ 1 & 0 & 0 & 0 \\ 0 & 1 & 0 & 0 \end{pmatrix}$$

*The equality*

$S_2(a,x) =$

$$\begin{pmatrix} (3-a)x^0+ax^0 & -(3-a)x^1-ax^1 & -(3-a)x^2 & -(3-a)x^3 \\ & -1+a-(a-1) & -ax^2+1+2 & -ax^3 \\ (3-a)x^1+ax^1 & (3-a)x^0+ax^0 & (3-a)x^3-ax^3 & -(3-a)x^2 \\ +1-a+a-1 & & & +ax^2-1+2 \\ (3-a)x^2 & -(3-a)x^3+ax^3 & (3-a)x^0+ax^0 & (3-a)x^1-ax^1 \\ +ax^2-1-2 & & & -1+a+a-1 \\ (3-a)x^3+ax^3 & (3-a)x^2 & -(3-a)x^1+ax^1 & (3-a)x^0+ax^0 \\ & -ax^2+1-2 & +1-a-(a-1) & \end{pmatrix}$$

*follows from equalities* (2.12.5), (2.12.6), (2.12.7), (2.12.8), (2.12.9). *Therefore*
$S_2(a,x) =$

(2.12.10)
$$\begin{pmatrix} 3x^0 & -3x^1 & -3x^2+3 & -3x^3 \\ 3x^1 & 3x^0 & (3-2a)x^3 & -(3-2a)x^2+1 \\ 3x^2-3 & 3x^3 & 3x^0 & 3x^1 \\ 3x^3 & (3-2a)x^2-1 & -(3-2a)x^1 & 3x^0 \\ & & -2(a-1) & \end{pmatrix}$$

*From the equality* (2.12.10) *it follows that* $x = j$, $a = 1$. *I recall that to reduce calculations I considered* $a \in R$. *In general, I had to consider* $a \in H$.  □

It is very important to note that even polynomial $p(x)$ has at least two roots, I found infinitely many quotients of polynomial $p(x)$ by polynomial $r(x)$. However, majority of quotients never equal 0, and only one case has a root the second root of a polynomial $p(x)$.

**October 10.** I was ready to start to answer on the question 2.12.2. However I have met very interesting question.

If $a \ne 1$, then $s_2(a,x) \ne 0$. However

(2.12.11) $$s_2(a,j) \circ r(j) = 0$$

From the equality (2.12.11) it follows that $s_2(a,j)$ is singular map and

$$r(j) \in \ker s_2(a,j)$$

I came close to answering the question 2.12.2. Because the answer to this question is similar to last remark.

Namely. Let for tensors $s_1(x), s_2(x) \in A \otimes A$ following equalities are true
$$p(x) = s_1(x) \circ r(x)$$
$$p(x) = s_2(x) \circ r(x)$$
Then
$$(s_1(x) - s_2(x)) \circ r(x) = 0$$
Therefore
$$r(x) \in \ker(s_1(x) - s_2(x))$$
Therefore, if I know at least one quotient $s_1(x)$, then any quotient $s_2(x)$ has form

(2.12.12) $$s_2(x) = s_1(x) + d(x)$$

where
$$r(x) \in \ker d(x)$$
Here I am confused that coefficients of the matrix $d(x)$ can be fractional linear functions $x$. However I have to consider this in an exemple.

And finally, there is interesting consequence of my remark. If I found $s_1(x)$, can I evaluate the possible roots of polynomial $p(x)$. Yes. These are the values of $x$ for which
$$p(x) \in \ker s_1(x)$$

DEFINITION 2.12.4. *We can prove that*
$$s_2(a, j) \circ r(j) = 0$$

*in two ways.*

- *Since $x = j$, then the eqaulity*

(2.12.13) $$s_2(a, j) = (3-a) \otimes j + (aj - j + i - ai) \otimes 1 + (a-1) \otimes i - 2 \otimes j$$

*follows from the eqaulity (2.12.4) and the eqaulity*

(2.12.14) $$r(j) = j - i$$

*follows from the eqaulity (2.12.1). Eqaulities*
$$s_2(a, j) \circ (j - i)$$
$$= (3-a)(j-i)j + (aj - j + i - ai)(j-i) + (a-1)(j-i)i - 2(j-i)j$$
$$= (3-a)(-1-k) + ((a-1)j + (1-a)i)(j-i)$$
$$+ (a-1)(-k+1) - 2(-1-k)$$

(2.12.15)
$$s_2(a, j) \circ (j-i) = -(a-1)(-1-k) + ((a-1)j + (1-a)i)j$$
$$- ((a-1)j + (1-a)i)i + (a-1)(-k+1)$$
$$= 2(a-1) - (a-1) + (1-a)k + (a-1)k + (1-a)$$
$$= 0$$

*follow from the eqaulities (2.12.13), (2.12.14).*

- From the eqaulity (2.12.10) it follows that matrix $S_2(a,j)$ has form

(2.12.16)
$$\begin{pmatrix} 0 & 0 & 0 & 0 \\ 0 & 0 & 0 & -(3-2a)+1 \\ 0 & 0 & 0 & 0 \\ 0 & (3-2a)-1 & -2(a-1) & 0 \end{pmatrix}$$

Therefore, the matrix $S_2(a,j)$ is singular when $x = j$ and

(2.12.17)
$$\begin{pmatrix} 0 & 0 & 0 & 0 \\ 0 & 0 & 0 & -2+2a \\ 0 & 0 & 0 & 0 \\ 0 & 2-2a & -2a+2 & 0 \end{pmatrix} \begin{pmatrix} 0 \\ -1 \\ 1 \\ 0 \end{pmatrix} = \begin{pmatrix} 0 \\ 0 \\ 0 \\ 0 \end{pmatrix}$$

$\square$

**October 11.** Now I am ready to return to polynomial

(2.12.18)
$$\begin{aligned} p(x) &= (-4 \otimes 1 \otimes 1 - 4 \otimes i \otimes i - 4 \otimes j \otimes j - 4 \otimes k \otimes k) \circ x^2 \\ &+ (4 \otimes i + 2j \otimes 1 - 2k \otimes i - 2 \otimes j + 2i \otimes k) \circ x + 1 + 2k \\ &= -4x^2 - 4xixi - 4xjxj - 4xkxk \\ &+ 4xi + 2jx - 2kxi - 2xj + 2ixk + 1 + 2k \end{aligned}$$

and I want to divide this polynomial to polynomial

(2.12.19)
$$q(x) = x - \frac{1}{4}(i+j)$$

The equality

(2.12.20)
$$4x = 4q(x) + i + j$$

follows from the equality (2.12.19). Equalities

(2.12.21)
$$\begin{aligned} p(x) &= -4q(x)(x + ixi + jxj + kxk - i) \\ &- (i+j)(x + ixi + jxj + kxk - i) \\ &+ 2jx - 2kxi - 2xj + 2ixk + 1 + 2k \\ &= -4q(x)(x + ixi + jxj + kxk - i) \\ &- i(x + ixi + jxj + kxk) \\ &- j(x + ixi + jxj + kxk) \\ &+ (i+j)i \\ &+ 2jx - 2kxi - 2xj + 2ixk + 1 + 2k \end{aligned}$$

## 2.12. Division of polynomials

$$p(x) = -4q(x)(x + ixi + jxj + kxk - i)$$
$$- ix + xi - kxj + jxk$$
$$- jx + kxi + xj - ixk$$
$$- 1 - k$$
$$+ 2jx - 2kxi - 2xj + 2ixk + 1 + 2k$$
$$= -4q(x)(x + ixi + jxj + kxk - i)$$
$$- ix + xi - kxj + jxk$$
$$+ jx - kxi - xj + ixk + k$$

$$p(x) = -4q(x)(x + ixi + jxj + kxk - i)$$
$$- i(q(x) + \frac{1}{4}i + \frac{1}{4}j) + (q(x) + \frac{1}{4}i + \frac{1}{4}j)i$$
$$- k(q(x) + \frac{1}{4}i + \frac{1}{4}j)j + j(q(x) + \frac{1}{4}i + \frac{1}{4}j)k$$
$$+ j(q(x) + \frac{1}{4}i + \frac{1}{4}j) - k(q(x) + \frac{1}{4}i + \frac{1}{4}j)i$$
$$- (q(x) + \frac{1}{4}i + \frac{1}{4}j)j + i(q(x) + \frac{1}{4}i + \frac{1}{4}j)k + k$$

$$p(x) = -4q(x)(x + ixi + jxj + kxk - i)$$
$$- iq(x) - \frac{1}{4}ii - \frac{1}{4}ij + q(x)i + \frac{1}{4}ii + \frac{1}{4}ji$$
$$- kq(x)j - \frac{1}{4}kij - \frac{1}{4}kjj + jq(x)k + \frac{1}{4}jik + \frac{1}{4}jjk$$
$$+ jq(x) + \frac{1}{4}ji + \frac{1}{4}jj - kq(x)i - \frac{1}{4}kii - \frac{1}{4}kji$$
$$- q(x)j - \frac{1}{4}ij - \frac{1}{4}jj + iq(x)k + \frac{1}{4}iik + \frac{1}{4}ijk + k$$
$$= -4q(x)(x + ixi + jxj + kxk - i)$$
$$- iq(x) + \frac{1}{4} - \frac{1}{4}k + q(x)i - \frac{1}{4} - \frac{1}{4}k$$
$$- kq(x)j + \frac{1}{4} + \frac{1}{4}k + jq(x)k + \frac{1}{4} - \frac{1}{4}k$$
$$+ jq(x) - \frac{1}{4}k - \frac{1}{4} - kq(x)i + \frac{1}{4}k - \frac{1}{4}$$
$$- q(x)j - \frac{1}{4}k + \frac{1}{4} + iq(x)k - \frac{1}{4}k - \frac{1}{4} + k$$

$$p(x) = -4q(x)(x + ixi + jxj + kxk - i)$$
$$- iq(x) + q(x)i - kq(x)j + jq(x)k$$
$$+ jq(x) - kq(x)i - q(x)j + iq(x)k$$

$$p(x) = -4q(x)(x + ixi + jxj + kxk - i)$$
$$- kq(x)(i + j) + (i + j)q(x)k$$
$$+ (j - i)q(x) + q(x)(i - j)$$
(2.12.22)
$$= (-4 \otimes (x + ixi + jxj + kxk - i)$$
$$- k \otimes (i + j) + (i + j) \otimes k$$
$$+ (j - i) \otimes 1 + 1 \otimes (i - j)) \circ q(x)$$

follow from equalities (2.12.18), (2.12.20). From the equality (2.12.22), it follows that we have to find the matrix

(2.12.23) $S(x) = 8E_r(x^* - i) - E_l(k)E_r(i+j) + E_l(i+j)E_r(k) + E_l(j-i) + E_r(i-j)$

and its kernel for

$$x = \frac{1}{4}(i + j)$$

Consider every term.

(2.12.24) $E_r(x^* - i) = \begin{pmatrix} x^0 & x^1 + 1 & x^2 & x^3 \\ -x^1 - 1 & x^0 & -x^3 & x^2 \\ -x^2 & x^3 & x^0 & -x^1 - 1 \\ -x^3 & -x^2 & x^1 + 1 & x^0 \end{pmatrix}$

(2.12.25) $E_l(k) = \begin{pmatrix} 0 & 0 & 0 & -1 \\ 0 & 0 & -1 & 0 \\ 0 & 1 & 0 & 0 \\ 1 & 0 & 0 & 0 \end{pmatrix}$

(2.12.26) $E_r(i+j) = \begin{pmatrix} 0 & -1 & -1 & 0 \\ 1 & 0 & 0 & -1 \\ 1 & 0 & 0 & 1 \\ 0 & 1 & -1 & 0 \end{pmatrix}$

(2.12.27) $E_l(k)E_r(i+j) = \begin{pmatrix} 0 & -1 & 1 & 0 \\ -1 & 0 & 0 & -1 \\ 1 & 0 & 0 & -1 \\ 0 & -1 & -1 & 0 \end{pmatrix}$

## 2.12. Division of polynomials

$$(2.12.28) \qquad E_l(i+j) = \begin{pmatrix} 0 & -1 & -1 & 0 \\ 1 & 0 & 0 & 1 \\ 1 & 0 & 0 & -1 \\ 0 & -1 & 1 & 0 \end{pmatrix}$$

$$(2.12.29) \qquad E_r(k) = \begin{pmatrix} 0 & 0 & 0 & -1 \\ 0 & 0 & 1 & 0 \\ 0 & -1 & 0 & 0 \\ 1 & 0 & 0 & 0 \end{pmatrix}$$

$$(2.12.30) \qquad E_l(i+j)E_r(k) = \begin{pmatrix} 0 & 1 & -1 & 0 \\ 1 & 0 & 0 & -1 \\ -1 & 0 & 0 & -1 \\ 0 & -1 & -1 & 0 \end{pmatrix}$$

$$(2.12.31) \qquad E_l(j-i) = \begin{pmatrix} 0 & 1 & -1 & 0 \\ -1 & 0 & 0 & 1 \\ 1 & 0 & 0 & 1 \\ 0 & -1 & -1 & 0 \end{pmatrix}$$

$$(2.12.32) \qquad E_r(i-j) = \begin{pmatrix} 0 & -1 & 1 & 0 \\ 1 & 0 & 0 & 1 \\ -1 & 0 & 0 & 1 \\ 0 & -1 & -1 & 0 \end{pmatrix}$$

$$- E_l(k)E_r(i+j) + E_l(i+j)E_r(k) + E_l(j-i) + E_r(i-j)$$

(2.12.33)
$$= \begin{pmatrix} 0 & 1+1+1-1 & -1-1-1+1 & 0 \\ 1+1-1+1 & 0 & 0 & 1-1+1+1 \\ -1-1+1-1 & 0 & 0 & 1-1+1+1 \\ 0 & 1-1-1-1 & 1-1-1-1 & 0 \end{pmatrix}$$

$$= \begin{pmatrix} 0 & 2 & -2 & 0 \\ 2 & 0 & 0 & 2 \\ -2 & 0 & 0 & 2 \\ 0 & -2 & -2 & 0 \end{pmatrix}$$

(2.12.34)
$$S(x) = \begin{pmatrix} 8x^0 & 8x^1+8+2 & 8x^2-2 & 8x^3 \\ -8x^1-8+2 & 8x^0 & -8x^3 & 8x^2+2 \\ -8x^2-2 & 8x^3 & 8x^0 & -8x^1-8+2 \\ -8x^3 & -8x^2-2 & 8x^1+8-2 & 8x^0 \end{pmatrix}$$

$$= \begin{pmatrix} 8x^0 & 8x^1+10 & 8x^2-2 & 8x^3 \\ -8x^1-6 & 8x^0 & -8x^3 & 8x^2+2 \\ -8x^2-2 & 8x^3 & 8x^0 & -8x^1-6 \\ -8x^3 & -8x^2-2 & 8x^1+6 & 8x^0 \end{pmatrix}$$

The equality

$$S(\frac{1}{4}(i+j)) = \begin{pmatrix} 0 & 2+10 & 2-2 & 0 \\ -2-6 & 0 & 0 & 2+2 \\ -2-2 & 0 & 0 & -2-6 \\ 0 & -2-2 & 2+6 & 0 \end{pmatrix}$$

$$= \begin{pmatrix} 0 & 12 & 0 & 0 \\ -8 & 0 & 0 & 4 \\ -4 & 0 & 0 & -8 \\ 0 & -4 & 8 & 0 \end{pmatrix}$$

follows from the equality (2.12.34). This is highly eloquent answer. This matrix is non singular. Therefore, the root

$$x = \frac{1}{4}(i+j)$$

has multiplicity 1.

I see that this method of studying the set of polynomial divisors of the polynomial is far from perfect. This is not simple task to search values $x$ for which the matrix (2.12.34) is not singular. However this is only first step. Actually, I solved more difficult task. I realized that this set may be infinite and that this set has well defined structure.

# References

[1] J. D. Anderson, P. A. Laing, E. L. Lau, A. S. Liu, M. M. Nieto, and S. G. Turyshev, Indication, from Pioneer 10/11, Galileo, and Ulysses Data, of an Apparent Anomalous, Weak, Long-Range Acceleration, Phys. Rev. Lett. 81, 2858, (1998), eprint arXiv:gr-qc/9808081 (1998)

[2] J. D. Anderson, P. A. Laing, E. L. Lau, A. S. Liu, M. M. Nieto, and S. G. Turyshev, Study of the anomalous acceleration of Pioneer 10 and 11, Phys. Rev. D 65, 082004, 50 pp., (2002), eprint arXiv:gr-qc/0104064 (2001)

[3] Serge Lang, Algebra, Springer, 2002

[4] P. K. Rashevsky, Riemann Geometry and Tensor Calculus, Moscow, Nauka, 1967

[5] C.A. Deavours, The Quaternion Calculus, American Mathematical Monthly, **80** (1973), pp. 995 - 1008

[6] Kevin McCrimmon; A Taste of Jordan Algebras; Springer, 2004

[7] Aleks Kleyn, Lectures on Linear Algebra over Division Ring, eprint arXiv:math.GM/0701238 (2010)

[8] Aleks Kleyn, Lorentz Transformation and General Covariance Principle, eprint arXiv:0803.3276 (2009)

[9] Aleks Kleyn, Introduction into Calculus over Division Ring, eprint arXiv:0812.4763 (2010)

[10] Aleks Kleyn, Representation of Universal Algebra, eprint arXiv:0912.3315 (2009)

[11] Aleks Kleyn, Linear Maps of Free Algebra, eprint arXiv:1003.1544 (2010)

[12] Aleks Kleyn, $C^*$-Rhapsody, eprint arXiv:1104.5197 (2011)

[13] Aleks Kleyn, Basis of Representation of Universal Algebra, eprint arXiv:1111.6035 (2011)

[14] Aleks Kleyn, Alexandre Laugier, Orthonormal Basis in Minkowski Space, eprint arXiv:1201.4158 (2012)

[15] Aleks Kleyn, Maps of Conjugation of Quaternion Algebra, eprint arXiv:1202.6021 (2012)

[16] Aleks Kleyn.
Single Variable Calculus: Noncomutative Banach Algebra.

CreateSpace Independent Publishing Platform, 2014;
ISBN-13: 978-1497563810

[17] Aleks Kleyn, Crash Course in Calculus over Banach Algebra
CreateSpace Independent Publishing Platform, 2018;
ISBN-13: 978-1985666931

[18] Aleks Kleyn, Quadratic Equation over Associative $D$-Algebra
Kindle Direct Publishing, 2018;
ISBN-13: 978-1728793399

[19] Paul M. Cohn, Universal Algebra, Springer, 1981

[20] Postnikov M. M., Geometry IV: Differential geometry, Moscow, Nauka, 1983

[21] A. Sudbery, Quaternionic Analysis,
Math. Proc. Camb. Phil. Soc. (1979), **85**, 199 - 225

[22] A. Sudbery, Quaternionic Analysis,
eprint ResearchGate:2657821 (1977)

# Index

associative $D$-algebra  29
associator of $D$-algebra  29
auto parallel line  44

coordinates of associator  30

extreme line  44

free representation  47, 48, 48, 52

metric-affine manifold  43

# Special Symbols and Notations

$(a, b, c)$  associator of $D$-algebra  29
$A_{ijl}^k$  coordinates of associator  30

www.ingramcontent.com/pod-product-compliance
Lightning Source LLC
Chambersburg PA
CBHW041314180526
45172CB00004B/1095